SOLUTION ARCHITECTURE FOUNDATIONS

BCS, THE CHARTERED INSTITUTE FOR IT

BCS, The Chartered Institute for IT, is committed to making IT good for society. We use the power of our network to bring about positive, tangible change. We champion the global IT profession and the interests of individuals, engaged in that profession, for the benefit of all.

Exchanging IT expertise and knowledge
The Institute fosters links between experts from industry, academia and business to promote new thinking, education and knowledge sharing.

Supporting practitioners
Through continuing professional development and a series of respected IT qualifications, the Institute seeks to promote professional practice tuned to the demands of business. It provides practical support and information services to its members and volunteer communities around the world.

Setting standards and frameworks
The Institute collaborates with government, industry and relevant bodies to establish good working practices, codes of conduct, skills frameworks and common standards. It also offers a range of consultancy services to employers to help them adopt best practice.

Become a member
Over 70,000 people including students, teachers, professionals and practitioners enjoy the benefits of BCS membership. These include access to an international community, invitations to a roster of local and national events, career development tools and a quarterly thought-leadership magazine. Visit www.bcs.org/membership to find out more.

Further information
BCS, The Chartered Institute for IT,
3 Newbridge Square,
Swindon, SN1 1BY, United Kingdom.
T +44 (0) 1793 417 417
(Monday to Friday, 09:00 to 17:00 UK time)
www.bcs.org/contact
http://shop.bcs.org/

SOLUTION ARCHITECTURE FOUNDATIONS

Mark Lovatt

Published by BCS Learning and Development Ltd, a wholly owned subsidiary of BCS, The Chartered Institute for IT, 3 Newbridge Square, Swindon, SN1 1BY, UK.
www.bcs.org

Paperback ISBN: 978-1-78017-5652
PDF ISBN: 978-1-78017-5669
ePUB ISBN: 978-1-78017-5676

Ebook available

British Cataloguing in Publication Data.
A CIP catalogue record for this book is available at the British Library.

Publisher's acknowledgements
Reviewer: Luke Butcher
Publisher: Ian Borthwick
Commissioning editor: Rebecca Youé
Production manager: Florence Leroy
Project manager: Sunrise Setting Ltd
Copy-editor: Moira Eagling
Proofreader: Annette Parkinson
Indexer: John Silvester
Cover design: Alex Wright
Cover image: istock/George Clerk
Typeset by Lapiz Digital Services, Chennai, India.

CONTENTS

FIGURES AND TABLES

AUTHOR

Mark Lovatt has worked with complex systems since the 1970s, initially supporting business applications based on IBM mainframes with flexible data analytics using mini and microcomputer components. With the introduction of the first generation of IBM PC, he focused on enabling direct access to view and manipulate live data by business managers to make businesses more responsive to market changes. During the 1980s, Mark worked with early adopters of networked operating systems for local and enterprise connectivity. Through the 1990s he provided consultancy and training services for enterprise server platforms from Microsoft, IBM and Sun Microsystems.

For the past 20 years Mark has specialised in large-scale healthcare systems, adapting and integrating proprietary systems and medical equipment to fit with service-based operating procedures at NHS hospitals, primary care and in the private sector.

Throughout a long and varied career, Mark's focus has been on the interface between business and IT to ensure that investments are made with realistic expectations and in alignment with corporate strategy. He has combined consultancy, education, software development and support working on major vendor-provided systems and open-source platforms.

Currently, Mark is deputy chief examiner for solution development for BCS and director of Mark Lovatt Associates, providing consultancy and training services in enterprise, data and solution architecture, and software development.

ABBREVIATIONS

5F	Five Forces
ABB	Architecture Building Block
AD	Architecture Description
ADM	Architecture Development Method
AES	Advanced Encryption Standard
AI	Artificial Intelligence
API	Application Programming Interface
APM	Association of Project Managers
BA	Business Analyst
BCS	BCS, The Chartered Institute for IT
BI	Business Intelligence
BM	Board Meeting
CAS	Clinic Administration System*
CDN	Content Delivery Network
CEO	Chief Executive Officer
CFO	Chief Finance Officer
CI/CD	Continuous Integration, Continuous Delivery
CMDB	Configuration Management Database
CMS	Communication Management System*
COBIT	Control Objectives for Information and Related Technologies
CRM	Customer Relationship Management
CRUD	Create, Read, Update, Delete
CSF	Critical Success Factor
CWE	Common Weakness Enumeration
DA	Design Authority
DevOps	Development Operations
DoDAF	Department of Defense Architecture Framework
DSDM	Dynamic Systems Development Method
DSS	Decision Support System
EA	Enterprise Architecture

ERD	Entity Relationship Diagram
ERM	Enterprise Risk Management
ERP	Enterprise Resource Planning
FD	Finance Director
FYE	Financial Year End
GB	Gigabit
GDPR	General Data Protection Regulations
GIS	Geographical Information System
GP	General Practitioner
GPS	Global Positioning System
HR	Human Resources
IAM	Identity and Access Management
ID	Identification
IEC	International Electrotechnical Commission
IEEE	Institute of Electrical and Electronics Engineers
IoT	Internet of Things
ISO	International Organisation for Standardisation
ISP	Internet Service Provider
IT	Information Technology
KPI	Key Performance Indicator
LAN	Local Area Network
MCMS	Message Content Management System*
MI	Management Information
MOST	Mission, Objectives, Strategy, Tactics
NFC	Near Field Communication
NFR	Non-Functional Requirement
NIST	National Institute of Standards and Technology
NoSQL	Not only SQL
OCR	Optical Character Recognition
OMR	Optical Mark Recognition
OWASP	Open Web Application Security Project
P3M	Project, Programme and Portfolio Management
PESTLE	Political, Economic, Socio-cultural, Technological, Legal, Environmental
PII	Personally Identifiable Information
PMO	Programme Management Office
POPIT	People, Organisation, Process, Information, Technology
PR	Public Relations
QA	Quality Assurance

QoS	Quality of Service
RACI	Responsible, Accountable, Consulted, Informed
RAG	Red, Amber, Green
RBAC	Role-Based Access Control
RDBMS	Relational Database Management System
REST	REpresentational State Transfer
RFID	Radio-Frequency Identification
ROI	Return on Investment
RPA	Robotic Process Automation
SaaS	Software as a Service
SBB	Solution Building Block
SBBI	Solution Building Block Interface
SLA	Service Level Agreement
SMART	Specific, Measurable, Actionable, Realistic, Time-bound
SME	Subject Matter Expert
SQL	Structured Query Language
TOGAF®	The Open Group Architecture Framework
TPS	Transactions Per Second
TRM	Technical Reference Model
UI	User Interface
UML	Unified Modelling Language
VMOST	Vision, Mission, Objectives, Strategy, Tactics
VoC	Voice of the Customer
VP	Vice President
VPN	Virtual Private Network
VSM	Value Stream Map or Mapping
WBS	Work Breakdown Structure
WiFi	Wireless Fidelity
xor	Exclusive Or

*Part of the case study, not a more widely used abbreviation

PREFACE

This book is an introduction to the discipline of solution architecture, which is a structured way to address a problem, risk or opportunity and produce an effective solution. It is called solution architecture because of the comprehensive and systematic approach that ensures the solution design truly addresses the problem and will work within the larger structure of the business.

The main intended audience for this book comprises those entering the field, having been recently appointed to a solution architecture role, and those already in post who may not have received any formal training and who are looking for guidance on how best to perform the role of solution architect. Others who may be interested in obtaining an understanding of solution architecture include those who are involved in specifying, designing and delivering solutions, such as developers, analysts, information specialists and those in other architecture roles.

Solutions are not the same as IT systems. Plenty of problems can be solved, risks avoided and opportunities seized by changes in staff behaviour or making better use of existing resources, for example. Many solutions do involve new or modified IT systems, but these are rarely able to solve the problem on their own. Large IT projects often fail because the systems do not fit with the ways of working or structure of the business. Successful solutions must therefore be designed using an architectural approach that includes all relevant aspects of business and technology that need to change.

The job title *Solution Architect* covers a wide range of roles, often focused on a specialist area such as software or cloud infrastructure. This book takes a broader view of solution architecture, but includes the involvement of other architecture disciplines in the production of a solution design.

There are not many books with 'solution architecture' in their title available at the time of writing, certainly compared with the number on the subjects of enterprise or software architecture. Those that are available tend to focus on aspects of implementing a solution through software or cloud components, for example, rather than the process of designing a suitable solution using architectural techniques, which is this book's rationale. A selected list may be found in the *Further Reading* section of this book.

The book is organised into eight chapters that build upon each other to give a comprehensive understanding of the theory and practice of solution architecture. No prior knowledge is assumed. Where existing concepts, frameworks or techniques are introduced, a reference is provided to the appropriate source material.

Throughout the book there is a case study based on the need of a fictional hospital to improve communications with its patients. This case study is used to illustrate the concepts and techniques presented. Activities are provided in every chapter, so that the reader can investigate concepts and try out the techniques described to get a better understanding of the subject matter, and these are often drawn from the case study. Review questions are provided at the end of each chapter as an opportunity for the reader to check their understanding.

Chapter 1 introduces the reader to the concepts of architecture in general, and solution architecture in particular. The idea of treating an organisation or enterprise as a system, with components and interactions between them, is brought into play as a fundamental part of the solution architecture approach. There is also an introduction to the types of component that form part of a solution. The role of a solution architect and the activities performed as part of that role are discussed, including the products and outcomes of the solution architecture process. The chapter concludes with a discussion of the rationale for using solution architecture, including its objectives and benefits and linking these to the importance of business and IT strategy as drivers for change.

Chapter 2 places solution architecture in the context of enterprise and other architectures. Business architecture is highlighted as a driver for change and also the location of key solution components such as business processes. The relationship between other domain architectures and solution architecture is described. Applications, data and information, and infrastructure architectures are covered in this category. There are also sections on the specialist domains of software architecture, which is often required to provide solution components, and security architecture, which has a wide-ranging impact on solution architecture by providing directives, designs and components.

Chapter 3 provides the details of a framework for performing solution architecture in an effective and efficient way, using a life cycle to organise activities into distinct phases. Each phase is described in detail, including the activities involved and artefacts produced or modified. The chapter concludes with a discussion of the benefits and challenges of using an architecture framework.

Chapter 4 itemises the various inputs that are required for the process of solution architecture to be successful. Useful inputs from internal sources such as strategy and enterprise architecture are itemised and described, along with the products of business analysis that may address internal or external drivers. Different types of requirements and constraints that may be derived from these inputs or come from specific stakeholder needs are categorised and the process for incorporating them into solution design is discussed.

Chapter 5 focuses on a specific technique – gap analysis – that is used throughout the solution architecture process to quantify the level of change and provide details of the components involved. Examples are given of where it can be used effectively in the life cycle and the steps involved are described with the various outputs that can be produced using this technique. The important transition from design to implementation, using gap analysis to provide the basis for the delivery roadmap, completes the chapter.

Chapter 6 covers the relationship between solution architecture and the stakeholders, which is the essential link that ensures the solution meets the needs of the business.

Stakeholders are categorised based on the type of involvement they have with solution design or governance. An important architectural technique – the use of viewpoints and views to manage stakeholders' concerns about the solution – is covered here, as well as the major interaction points during the life cycle, such as defining solution scope and business case option selection.

Chapter 7 looks at the critical activity of solution technology definition, which is the main interface between solution architecture and infrastructure architecture and is where the physical implementation of the logical solution design is considered in detail. This activity is a process that involves a number of steps, which are detailed here, along with the inputs required and outputs produced. There is also a discussion of the decisions that flow from this process and their impact on the implementation of the solution.

Chapter 8 follows the steps that take place during implementation that solution architecture is directly involved with, either during the handover of the design to the delivery teams or in ensuring the governance of the delivery of solution components and the implementation of the solution for use by the business. Support for change and maintenance to the solution after implementation and ongoing communication with the business are also discussed.

1 INTRODUCTION TO SOLUTION ARCHITECTURE

This book is a foundation-level introduction to the discipline of solution architecture which uses a holistic approach to analyse problems and design solutions using the best available evidence from all relevant sources.

LEARNING OUTCOMES

When you have completed this chapter, you should be able to demonstrate an understanding of the following:

- Key architecture concepts
- Solution architecture as a discipline
- Viewing a business or organisation as a system
- Aligning solutions with business and IT strategy
- Typical activities involved in solution architecture
- The outcomes of solution architecture
- The solution architect's role
- The components that make up a solution
- The objectives and benefits of taking a solution architecture approach

1.1 ARCHITECTURE

An architectural approach to solving problems involves:

- Seeing the 'big picture' of the problem, including how it may be solved **conceptually**.
- Breaking the problem down into components and making models to see how they work together **logically** to solve the problem.
- Itemising the **physical** changes that are required to move from problem to solution, possibly in incremental stages.

This approach is strategic, holistic and progressive and is an alternative to applying tactical 'quick fixes' that are short-term and narrowly focused. Even if a tactical solution is required as an urgent response to a factor beyond the control of the organisation, this can be fitted into a schedule that leads to a longer-term solution. The worst outcome is that a short-term fix is put in place that subsequently needs to be undone in order to continue with the strategy. This is wasteful of resources and is likely to delay progress. Having a target architecture means that tactical solutions can be introduced as part of an intermediate architecture that has a defined way forward.

Anything beyond a limited, local change will benefit from an architectural approach that addresses as much of the organisation and its environment as is needed to solve the problem. Note that **problem** includes business **issues** and **risks** as well as **opportunities** that might otherwise be missed.

There exist many different definitions of architecture that describe both the process of architectural design and the essence of that design which must be represented using models and descriptions. Architecture depends on treating businesses, enterprises and solutions as systems with components that interact with each other to exhibit functionality. The following definition is adapted from TOGAF 9.2 (TOGAF 9.2, 2018) and ISO 42010 (ISO/IEC/IEEE 42010:2011, 2011):

Architecture: the **inherent structure** and **behaviour** of a system which may be present as a result of **design** and/or **evolution**.

- **Structure** refers to the components or **building blocks** from which the system is composed and how they are connected together.
- **Behaviour** refers to the effects of the operation of the system.
- **Design** refers to proactive decisions made to improve the business value of a system.
- **Evolution** refers to the reactive adaptation of a system in response to its use within its operating environment.
- The fact that architecture is **inherent** means that all systems have an architecture even if it has not been analysed, designed or documented.

The term architecture may also be used to mean the documentation describing the structure and behaviour of a system. This documentation is more formally known as an **architecture description (AD)**.

Architecture is also used to mean the discipline or set of activities concerned with the analysis and design of systems and the production of architecture descriptions.

ISO 42010

ISO 42010 is a standard that addresses the creation, analysis and sustainment of architectures of systems through the use of architecture descriptions.

The ISO 42010 standard defines a number of terms:

- **Architecture**: fundamental concepts or properties of a system in its environment embodied in its elements, relationships, and in the principles of its design and evolution.

- **Architecture description**: work product used to express an architecture.

- **Architecting**: process of conceiving, defining, expressing, documenting, communicating, certifying proper implementation of, maintaining and improving an architecture throughout a system's life cycle.

The **scope** of architectural activity (see Figure 1.1) and architecture descriptions may include:

- an **enterprise** such as a business or government department;

- a **specialist domain** within an enterprise such as **security**, **business**, **data**, **applications**, or **infrastructure**;

- an **information system**, **software** system, or a system **component**;

- a **solution** to a business problem or opportunity.

Figure 1.1 Scope of architectural activity

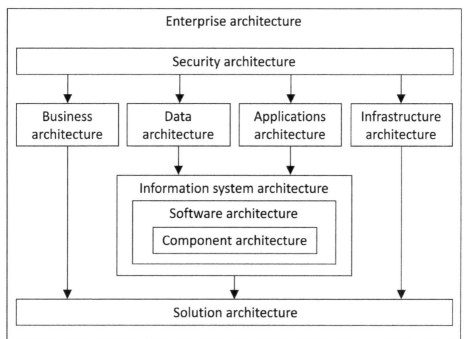

Models are used extensively in architecture. A model is an abstraction of something more complex. Architecture models are used to represent systems in the real world. Such models may be used to understand the current state of the system and to assess the effect of changes. Abstraction is used to simplify models using generalisation, composition and idealisation.

Architecture descriptions and models may represent both current and future states to demonstrate the scope and effect of change.

1.2 SOLUTION ARCHITECTURE

Solution architecture is a discipline concerned with the **production and management** of a **blueprint** for a comprehensive **solution,** that **addresses** a **business need,** **problem** or **opportunity**, and integrates with the business, in **alignment** with its **strategy**, while minimising **negative impacts**.

- **Blueprint** refers to a plan or outline (high-level design and definition of the structure and behaviour) of a solution to be put in place.
- **Production and management** refer to the dual responsibilities of solution architecture to coordinate the design activities whilst ensuring that changes are tested, validated and agreed at every stage.
- **Solutions** address **business needs**, **problems** or **opportunities** to improve the operation of a business system by, for example, increasing performance efficiency or reducing risk.
- **Negative impacts** must be minimised and any that cannot be eliminated must be identified so that their effects can be mitigated.
- On the positive side, solutions must contribute to the success of the business and therefore **align** with the business **strategy**.

What is a solution?

A solution addresses a problem, risk or opportunity facing a business. Over time many solutions are produced, even if they are not named as such, and once established they become indistinguishable from each other and part of the fabric of the enterprise. Solutions may address newly arising situations or seek improvements to the business by modifying existing provision in a specific area, in other words replacing an existing solution.

In the modern age, most solutions contain some technology, but it is important to recognise that other elements such as people and information are required to make them work. A solution therefore is holistic, considering all relevant aspects of the situation to find the best possible outcome for the business.

Solution architecture treats a solution as a system with component parts that interact to produce the required behaviour. The components may be new or existing and may be shared with other solutions. The interactions occur via interfaces between the components where information is exchanged.

Solution components may be classified into a checklist of five areas – people, organisational structures, processes, information and technology – to ensure they are not overlooked in the design.

To illustrate the concepts presented here, a case study will be used. The Fallowdale Hospital scenario (as described in the box below) will provide examples throughout this book to support the various aspects of solution architecture as they arise. The main objective in the scenario is to address the problems associated with the way the hospital communicates with patients.

Fallowdale Hospital case study

Fallowdale District Hospital is a general hospital providing a range of healthcare services to the medium-sized town of the same name and its surrounding district.

The majority of patients are referred to the hospital by a doctor, such as a general practitioner (GP) in their local surgery or health centre. Patients are then invited to attend for an appointment with a health professional, such as a specialist doctor, nurse or therapist, or for a diagnostic test or scan. Patients can also attend without an appointment at the emergency department and some other walk-in services.

The hospital currently communicates with all patients by letter to offer or confirm appointments and to provide results of tests and other outcomes of attendance. Letters are also sent when the patient or hospital cancels or reschedules an appointment.

A number of issues with patient communication have been identified:

- High cost of sending letters.
- High rate of non-attendance at appointments.
- Confidentiality risks of using the postal system.

Three separate business problems have been identified, although they may be related, and the hospital can therefore benefit from taking an architectural approach to addressing them. Some of the key aspects are:

- **Problem statement:** focused on a single issue (communicating with patients), the business problems are cost, missed appointments and confidentiality.
- **Blueprint:** the action plan for achieving the improvement they are seeking in patient communications.
- **Side effects:** imperative that the solution has zero negative impact elsewhere in the hospital.
- **Strategy:** the solution must support and ideally enhance the business and IT strategy.

1.3 THREE LEVELS OF A BUSINESS OR ORGANISATION

A business or any other organisation can be viewed as a system and modelled at three levels: business system, information system and IT system (see Figure 1.2).

Figure 1.2 Three levels of a business system

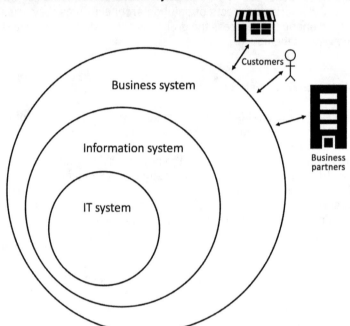

A **business system** is an organisation (or organisational unit) treated as a system with its own internal structure and behaviour, typically presenting services through external interfaces to customers and partner organisations.

Note the word 'business' in this context does not imply a commercial organisation but applies to any organisation or enterprise, representing what it does to achieve its aims and objectives.

An **information system** is a hybrid of human-managed processes and computer systems that manages business information. It may incorporate a range of information technology systems as well as manual activities within the context of a business system.

An **IT system** is typically a combination of hardware and software that is often used by an organisation to handle data and information within the context of an information system.

Fallowdale Hospital business system

The hospital operates a **business system** to enable patient referrals from GPs. This is offered as a business service to GP surgeries and local health authorities.

The referral system is based around an **information system** where many pieces of data and information are exchanged between people and sometimes with a computer system. Many exchanges are largely verbal, such as the initial assessment by the GP and appointments between specialists and the patient. Some information is recorded and transmitted on paper.

Parts of the referral system involve an **IT system**. These include the transmission of referral requests from the GP to the hospital, the schedule of hospital appointments and the healthcare professionals involved, and a record of any diagnoses, prescriptions and ongoing management of patient care.

Solution architecture is concerned with solutions that represent business systems that include within them information systems and IT systems. Some solutions have as their focus a new or modified information or IT system, but it is extremely rare for there to be no wider business involvement or impact.

1.4 ALIGNING SOLUTIONS WITH BUSINESS AND IT STRATEGY

Solution architecture must be aligned with business and IT strategy. VMOST analysis (Sondhi, 1999) is a technique for strategic planning that can be used to define strategies, refine strategic ideas and promote a consistent approach at multiple levels within the business (see Figure 1.3). There are many variations of VMOST, including a simplified version that does not include vision as a separate entity, called MOST (Cadle and Paul, 2021). VMOST is the most widely used of these techniques.

Figure 1.3 VMOST analysis

1.4.1 Vision

This represents the ideal future situation. Vision can be at multiple levels. There can be an overall vision for the organisation or enterprise and separate ones for specific business areas. There is often a vision for a problem area that can become the vision for a solution. A vision does not have to be achievable, or at least it does not have to be clear how it can be achieved. The vision just needs to be something that everyone can agree is how they would like things to be.

For example, part of the vision of a supermarket company might be that people are excited to shop there. This gives everyone a mental picture, and this is often the basis for the carefully selected images that are presented in advertisements.

The vision is closely linked to the mission, but will change as the organisation makes progress and the environment in which it operates changes.

Fallowdale Hospital vision

The board of the hospital has identified the following vision which it hopes will become reality within the next five years:

'We treat every person as an individual and provide the very best service available, tailored to the individual.'

1.4.2 Mission

This is the enduring concept behind the organisation. In a fast-changing world, even the mission may change but, if this happens, it probably indicates that something fundamental has changed about the organisation.

An organisation whose mission has probably not changed since its founding is the World Wildlife Fund. The current stated mission is 'to create a world where people and wildlife can thrive together' (World Wildlife Fund, 2021). This clarifies the point that humanity exists as part of nature and individual species cannot be conserved in isolation from the rest of the environment.

Fallowdale Hospital mission

'To improve the health of our patients and communities through delivering the best in clinical care, research, innovation and education.'

1.4.3 Objective

This is a target that must be met to achieve the vision and mission.

Multiple objectives are usually needed for any significant change, such as the implementation of a new solution, and their achievement marks the progression towards change.

A useful checklist for objectives is **SMART**. A SMART objective must be Specific, Measurable, Actionable, Realistic and Time-bound (Doran, 1981).

- **Specific:** to the organisation and situation so stakeholders can understand it and determine when it has been achieved.

- **Measurable:** so stakeholders definitively know whether it has been met (or partially met); sometimes expressed using **key performance indicators (KPIs)** and **critical success factors (CSFs)**.

- **Actionable:** alternatives such as *attainable* or *achievable* overlap with *realistic*, so *actionable* is preferred as the organisation must believe that action can achieve the outcome.

- **Realistic:** otherwise no one will take it seriously. Judgement can be based on past experience or modelling the future to assess the proposed actions.

- **Time-bound:** internal or external deadlines or the dependency of another piece of work may constrain its achievement, or a purely project-based time limit may be imposed.

Fallowdale Hospital objectives

Two of the key objectives that have been identified as necessary to the achievement of the vision are:

- Hospital staff must have access to all relevant information about the patient or service user at the time when it is needed to provide care. Each type of care offered must have a specification of what information is required by each party, including the patient, and when it is needed in the episode of care. A specification must be in place before care can be offered and must be reviewed at least annually.

- Patients must have access to relevant information in sufficient time before receiving care. Delivery of information from the hospital must be guaranteed with acknowledgement by the patient if required in the specification.

Activity 1.1

An internet service provider is looking for a solution to providing help and assistance to existing and new clients. The company is fairly successful, but it has a mixed reputation for customer service. No decision has been made as to what type of solution it will choose. It is a public company with shareholders and is listed on the stock exchange.

Create a SMART objective for the proposed solution based on the business problem identified.

1.4.4 Strategy

Strategy describes what an organisation or individual intends to do to achieve its objectives.

Strategy: detailed plan for achieving success in situations such as war, politics, business, industry, or sport, or the skill of planning for such situations.
(Cambridge Dictionary, 2021)

An organisation may have multiple strategies at different levels. At the top level are those for business and IT that apply to the entire organisation or enterprise. At lower levels, there may be sub-strategies that apply to parts of the business such as divisions and departments and initiatives such as geographical expansion.

Strategic decisions must be recorded and communicated to relevant stakeholders within the organisation whilst acknowledging that parts of the strategy may be confidential.

Each strategy should be consistent with itself and other strategies. To be consistent here means that there is no conflict between strategies; even if they are trying to achieve different things, they should not work against each other.

Strategies must aim to support the organisation's mission and vision and achieve its stated objectives. These are the measures by which the success of a strategy can be measured.

Strategies can change, especially if they are not succeeding, but also if circumstances change. All strategies should be reviewed regularly and strategic decisions communicated to all relevant stakeholders.

Fallowdale Hospital strategy

The main strategic decision that the board has agreed on is to dramatically improve communication between the hospital and those who use its services. On reflection, it was agreed that the provision of care is also dependent on efficient and effective communication between those providing the care.

The **business aspects** of this strategy are to:

- communicate in a consistent way, irrespective of source or destination;
- focus on communication that is relevant to the provision of care.

The **technological aspects** of this strategy are to:

- automate communication wherever appropriate;
- use existing systems and devices if possible;
- use communication technologies that are appropriate and compatible with their users.

1.4.5 Tactics

A tactic is an action that has been identified as an appropriate way to achieve one or more strategic outcomes. Each tactic only addresses part of a strategy and therefore multiple individual tactical actions are required to achieve all the strategic aims.

Since tactics are smaller units of work, they each involve less effort, money and time than the strategy they support. This means it is easier to modify or abandon a tactic if it is judged to be ineffective. However, tactical changes can have side effects and so it is critical to assess the impact of all changes to the approach being taken.

Fallowdale Hospital tactics

The tactics that have been suggested so far are shown in the following table:

Strategy	Tactic
Unify communication	New solution for patient/service user communication
Information relevant to care	Requirement for new solution
Maximum automation	Requirement for new solution
Use existing technology	Constraint for new solution
Encompass all users	Constraint for new solution

1.5 TYPICAL ACTIVITIES INVOLVED IN SOLUTION ARCHITECTURE

A considerable amount of work is required to achieve a successful solution, which typically starts very simply with an idea of how things could be better and must then be developed into a business change that can be put in place.

This work is carried out or governed by various people in the business, collectively known as stakeholders.

Stakeholder: ISO 42010 defines stakeholder as an 'individual, team, organisation, or classes thereof, having an interest in a system'.

(ISO/IEC/IEEE 42010:2011, 2011)

At the centre of these activities is the solution architecture function, typically led by a solution architect. Complex solutions may be broken down into smaller units, with each one being led by a solution architect and the overall solution coordinated by a senior solution architect.

The activities include:

- organising the process;
- interacting with stakeholders;
- building models;
- analysing the problem and recommending solution options;
- advising the business about the solution;
- governing the delivery.

1.5.1 Organising the process

The solution architecture function has some responsibilities that overlap with those of a project manager. In fact, the solution architecture process or life cycle is very much like a project or programme in that it has defined objectives and is composed of tasks and dependencies and is subject to the constraints of time, scope and resources. The difference is that solution architecture is concerned with designing and elaborating a solution, whereas project or programme management, in this context, is concerned with delivering the solution that has already been designed.

There is nothing to stop the solution architecture function from using the facilities provided by a programme management office (PMO) to organise meetings and manage communication, but all the activities are under the control of solution architecture.

An important aspect of organisation is managing timescales. This includes:

- Estimating and setting the duration of phases; for example, the sign-off by the business to start the solution architecture process is based on an estimate of the time and resources that will be required to complete the discovery phase.
- Agreeing dates and sequences of events, such as meetings, deliverables and who will review them.
- Handling any delays, for example to the production of artefacts.

Another aspect that requires a degree of control is the scope of the solution and of the architecture work required to deliver its design. Controlling scope is made easier if the problem or opportunity being addressed by the solution is clearly defined.

1.5.2 Interacting with other stakeholders

Each individual stakeholder must by definition have an interest in or concern with the solution. Stakeholders as a group must represent the business through the specification, design and delivery of the solution. Solution architecture is responsible for ensuring that the correct stakeholders have been identified. If the stakeholders are not fully representative of the business, there is a risk that the solution will not be fit for purpose. This could be because not all business needs were identified, or that there is a negative impact on business services in an unrepresented part of the business.

Solution architecture is also responsible for facilitating appropriate communication with and between stakeholders. Stakeholder communication is required for the following purposes:

- **Information:** important for transparency and keeping everyone aware of developments; usually one-way, but can result in requests for clarification and requests for closer involvement.

- **Consultation:** captures business expertise relating to the solution and allows the best-placed stakeholders to make fully informed decisions; this is two-way and proper records are essential for traceability.

- **Accountability:** records the approval of decisions and actions by specific stakeholders and may depend on authenticated and verifiable signatures.

- **Responsibility:** allocates and tracks tasks that are given to individuals and teams to manage and complete, including constraints and deadlines; interaction may be between stakeholders rather than via solution architecture.

These types of communication are part of the **RACI model** (Jacka and Keller, 2009), which is often represented as a matrix as a way of organising stakeholders and activities (see Figure 1.4).

A RACI matrix has rows representing activities and columns representing roles. The cell at the intersections of a row and a column is used to indicate the type of involvement of the role in the activity. An 'R' is used where the role is **responsible** for performing the activity. An 'A' indicates **accountability** for the success of the activity and is nearly always assigned to a single role. 'C' indicates that the role is **consulted** during the activity and 'I' means roles are **informed** of its progress and outcome.

Figure 1.4 RACI matrix

	Stakeholder group 1	Stakeholder group 2	Stakeholder group 3	Stakeholder group 4	Stakeholder group 5	Stakeholder group 6
Activity 1	R	I		A		I
Activity 2	C	R	I	A		I
Activity 3	R	R		I		A
Activity 4			A	I		I
Activity 5	I	C	R	A	C	

A RACI matrix is useful for summarising the involvement of stakeholders and others involved in architecture or project work. It also acts as a checklist to ensure all roles have been included. If no role has been made responsible for an activity, for example, it is unlikely to happen. If no role is to be consulted then it must be true that the responsible role has all the necessary information to perform the activity, and so on.

A good way to structure communication with stakeholders is to create a **design authority (DA)** for the solution. A design authority ensures the solution is aligned with the business by making decisions in accordance with its terms of reference as well as facilitating discussion and promoting understanding among stakeholders.

Design authorities are given responsibility for completing specific tasks and contribute to the overall governance of the design process.

Governance of solution architecture

For any type of organisational change, including designing and implementing a solution, governance means controlling activity and decision making to ensure that the change delivered matches the specification agreed with the business. The control required is achieved through the use of processes and organisational structures that provide a framework of authority and accountability within which the change can be delivered.

For solution architecture, the business is represented by stakeholders who agree the specification for the solution and are also involved in the governance of its delivery.

Governance must include assuring that products that are delivered match the specification but also account for the fact that the specification may need to be refined or altered during the delivery of the solution. This could be to fill in details that were not originally specified or because of changed circumstances in the business.

A key stakeholder role for solution architecture is the **business sponsor** who is ultimately accountable for the success of the solution. This must be a senior manager or director of the organisation, as they need sufficient authority over the solution area within the business to command the necessary finance and resources.

The business sponsor normally has too much responsibility within the business to be intimately involved in the details of governance and so may delegate much of the work to a body such as a **design authority**. This is constituted to have sufficient expertise and representation from the business to make the majority of decisions and escalate those it cannot to the business sponsor, who otherwise receives status updates and progress reports.

The design authority for a solution can be constituted and managed by a solution architect who can delegate tasks and handle queries. The design authority can keep other stakeholders informed with regular updates and reports and can seek expertise where it is beyond the knowledge of its members.

The requirements of governance vary between design and delivery of a solution. During design, decisions are concerned with building a specification for the solution that meets the needs of the business. During delivery, governance is mainly to check that what was specified is being delivered but also to make decisions when there is any variation from the specification.

Governance of solution architecture is successful when it ensures that what is designed and specified meets the needs of the business and that what is delivered matches the specification. The business also requires that resources are used efficiently and effectively, and that operational disruption is minimised. Governance must also ensure that these needs are met.

Activity 1.2

The Fallowdale Hospital board has the following directors:

- Chief executive officer (CEO)
- Chair
- Director of strategy
- Finance director (FD), in charge of IT
- Operations director, in charge of administrative staff in all areas including Outpatients
- Medical director, in charge of all doctors and medical specialties
- Director of nursing, in charge of inpatient, outpatient and specialist nurses

Using this information, draw up a list of potential stakeholders for the patient communications solution and, if possible, identify a business sponsor.

1.5.3 Building models

Models are central to architecture as they provide an interface between the future design of a system or solution that does not yet exist and the human understanding or vision of that future state (see Figure 1.5).

Models are critical in three other ways as they:

- enable understanding and analysis of the current state in the problem area;
- promote clear communication of ideas about the current and future states;
- demonstrate how the solution contained in the future state addresses the problem.

Figure 1.5 The role of the model in solution architecture

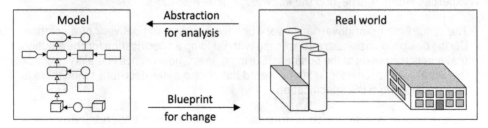

The solution architecture function is responsible for producing models of many types, which form part of artefacts that are used at the input, output or intermediate stages towards the selection, design and delivery planning of a solution.

Existing models, such as the current version of the solution, may need modification to bring them up to date or to focus on areas of interest.

New models are frequently required for communicating ideas with stakeholders to:

- elicit feedback;
- facilitate decision making;
- illustrate aspects of a solution;
- provide a contract with stakeholders about their concerns;
- represent the final design blueprint.

1.5.4 Analysing and selecting

Much of architectural design relies on:

- breaking down the current situation;
- analysing problems and opportunities;
- identifying components to keep;
- designing or specifying new components.

Given the amount of information available, a key competency for a solution architect is to be able to filter out what is irrelevant and focus on areas that need attention.

1.5.6 Providing specialist advice

The solution architecture function has the best knowledge and understanding of the solution, as is to be expected, but this extends far beyond the design of the final solution architecture to include detailed knowledge of:

- stakeholders and their concerns;
- initial and revised requirements;

- solution component specifications;
- enterprise architecture directives and constraints.

In short, the solution architecture function can legitimately be consulted about any of the following aspects of the proposed solution.

- **Explanations:** of stakeholder concerns, requirements or design decisions that are of particular interest to the business sponsor as well as other stakeholders for the current solution.
- **Details:** of models and other artefacts that may be of interest to other architects such as enterprise, data etc.
- **Discussions:** of the scope and ambition of the solution being designed that will inform those working on other solutions and projects within the enterprise.
- **Clarification:** of the outputs of the solution architecture process that are being used as a blueprint for building, implementing and deploying the solution.

1.5.7 Participating in governance of the delivery of the solution

Apart from being a point of consultation for architectural aspects of the solution, the solution architecture function is usually called upon to play a more formal governance role and to sign off plans and work products during the phases following the completion of the solution design.

- **Roadmap:** the point of handover of the delivery of the solution as a project and where solution architecture has the most influence over project goals and requirements. The solution architect or another member of the team may retain membership of the project or programme board to strengthen governance, tighten the alignment with the business requirements and promote capability maturity.
- **Solution development:** many critical decisions are made during the solution development life cycle. Solution architecture may be consulted and provide expert advice at any stage, but should be part of any governance structure such as a design authority. Any changes to the design that are made during this phase of implementation must be signed off by the business sponsor in consultation with solution architecture to maintain good governance.
- **Operations:** governance of the new solution when operational is the responsibility of the business area that requested it. Solution architecture can help with the definition of the SLAs and KPIs that are used to monitor the successful operation of the solution.
- **Benefits realisation:** this is the final part of the business change life cycle and focuses on measuring business benefits that were identified by or in collaboration with business analysis. Solution architecture has good insight into the traceability of the solution design back to the business requirements.
- **Change control and management:** ideally formal change control procedures should be in place for all solutions, so any future modifications to the solution or side impacts from changes elsewhere will be referred back to solution architecture to maintain existing solution benefits and minimise any negative impacts.

1.6 WHAT SOLUTION ARCHITECTURE AIMS TO DELIVER

The primary outputs of solution architecture are:

- a design with sufficient detail to be implemented;
- a plan or roadmap for the delivery of the solution to the business;
- an estimate of the cost of technological change.

Many supporting documents, such as reports and models, will be used and produced during solution architecture and these are collectively known as **artefacts**. They fall into three categories: inputs, deliverables (outputs) and intermediate work products (see Figure 1.6).

Figure 1.6 Solution architecture inputs, intermediate artefacts and outputs

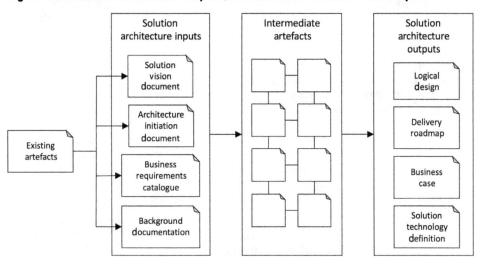

1.6.1 Inputs

Artefacts in this category are usually available before any solution architecture work takes place, although some may only become apparent after some discovery work has been done and others may need to be modified to be in a fit state to be used. Inputs are sometimes static but more commonly they are used, modified and transformed into the products of the solution architecture process.

The availability and quality of inputs to solution architecture vary according to the source of the solution concept.

The original idea for the solution may have arisen as a result of formal analysis, such as the business change life cycle from business analysis or the activities of enterprise architecture. If so, many of the inputs will be well-developed and additional background information is likely to be available.

Often, however, little or no analysis has occurred by the time the need for action has been identified. This means more work for the solution architecture team and other stakeholders during the discovery phase. Some artefacts are produced in order to start the process.

- **Solution vision statement:** also known as a solution concept definition, records that the organisation has identified the need for action and what the solution might look like – albeit abstract and conceptual with as little detail as possible about how the problem is to be solved; records a potential owner and other stakeholders.

- **Architecture initiation document:** gives authority to the solution architecture team to proceed to the next stage (discovery), estimates resource and time frame and lists deliverables.

- **Business requirements catalogue:** an initial collection of high-level business requirements directly linked to the problem; more will be identified during the discovery phase.

Other baseline artefacts may already exist.

Activity 1.3

The business requirements catalogue is an important input artefact to the solution architecture process. From the information given so far, find two or three high-level business requirements for the Fallowdale Hospital patient communications solution.

The following baseline architecture artefacts may exist or may need to be created or updated before being used as inputs:

- **Solution architecture:** artefacts for solutions being modified and overlapping with the solution in focus.

- **Enterprise architecture:** principles, policies and business rules.

- **Data architecture:** artefacts for data or information that will be used, managed or impacted by the proposed solution.

- **Business architecture:** process models that will be impacted by the proposed solution.

- **Applications architecture:** artefacts that may be a source of system building blocks.

- **Infrastructure architecture:** artefacts that may be a source of technical building blocks.

Input artefacts are useful at different points in solution architecture but the earlier they can be identified as required the better, as this allows time for creating missing items and updating any that are not sufficiently current.

1.6.2 Deliverables

These are the outputs of the solution architecture process that are used by other parts of the business.

- **Logical solution design:** a detailed design of the proposed solution that includes all components and interfaces that are involved in delivering the solution. This is a future state design with all new and existing components and the interfaces that connect them.

- **Delivery roadmap:** a high-level plan for the delivery of the changes required to implement and deploy the solution. This is based on a **gap analysis** between the current and future state models of the problem area and the solution.

Gap analysis is the systematic identification of differences or gaps between two architectures. The gaps indicate which parts of the first architecture need to change to achieve the transformation to the second architecture. This technique is covered in depth in Chapter 5.

- **Business case:** a record of the decisions made by the business relating to the solution, including consideration of alternatives.

- **Solution technology definition:** a list of IT and other infrastructure items that will be part of the solution, with an indication of where new items will be required or extra capacity for existing ones.

1.6.3 Intermediate work products

Many artefacts are produced and modified during the solution architecture process. These are working documents and are subject to change during one or more of the phases in the life cycle. Some artefacts provide direct inputs to other artefacts or are the basis for activities within the various life cycle phases.

- **Baseline architecture catalogue:** a categorised list of architecture artefacts that are available and that represent the position before the solution is implemented and deployed.

- **Solution stakeholder register:** a list of stakeholder categories with any associated concerns.

- **Stakeholder communication plan:** a list of decisions about how to engage with the stakeholders in the register. This will often include a **RACI matrix** as this is key to the type of interaction that will be required.

- **Solution outline definition:** a high-level conceptual outline of the type of solution being considered. This often has multiple options for the approach, scale and scope of the solution, which will need further investigation before further decisions are made by stakeholders representing the business.

- **Business case with multiple options:** a formal document justifying the solution based on its benefits and including details of the financial, human and other resources that will be dedicated to achieving it. It also serves as a record of the decision-making process and rationale for the choice of (usually) one solution to take forward.

- **Solution building block model:** a list of building blocks (components) that are expected to be part of the solution, together with some details about the part they will play.

- **Solution interface catalogue:** a list of interactions between components in the solution building block model and details of their characteristics, especially information exchange.

- **Viewpoints and views:** viewpoints are specifications of models that can be produced, and views are the models that are produced. Views are intended to address the concerns of stakeholders by showing specific aspects of the solution and affected areas of the business.

- **Model test reports:** models may be tested against scenarios to see the behaviour of one or more components of a solution under realistic conditions; these are the records of those tests.

- **Impact report:** the result of an impact analysis, including an assessment of all affected areas and the nature and extent of the impact.

- **Gap report:** the result of a gap analysis that compares the current and future state models of a solution and itemises the changes required to move to the future state. Usually includes a breakdown of the costs, resources and time that will need to be invested to achieve the change.

- **Gap models:** while performing a gap analysis, various models will be used, some produced specifically for the purpose of gap analysis. These are often marked up to show the changes and are a useful reference point for producing artefacts such as the delivery roadmap.

1.7 THE SOLUTION ARCHITECT'S ROLE

Typically, each proposed solution is the responsibility of a single solution architect. Very large problem areas should be broken down into realistically sized ones. The risk of accepting a very large problem area as a solution architect is that it cannot be solved as a unit.

Skills and knowledge areas that are required as a solution architect are:

- **Business:**
  - Business strategy.
  - Current/planned business model, architecture and operating model.
  - Organisational politics.
  - Commercial and market factors.
  - Trends in technology and business practice.
  - The operating environment as embodied in PESTLE (political, economic, sociocultural, technological, legal and environmental).

- **Technical:**
  - Technical strategy.
  - Current and planned infrastructure/technology architecture.
  - Methodological approaches to solution development.
- **Delivery:**
  - Project, portfolio and programme management (P3M).
  - Active and planned programmes and projects.
  - Change and configuration management.
  - Risk management.
- **Practical:**
  - Communication skills.
  - Stakeholder management.
  - Requirements handling.
  - Problem solving.
  - Innovation.
  - Leadership.

Some of these are considered technical and some as 'soft skills', but all are required in some measure to do the job of solution architect effectively.

The balance of skills required by a solution architect varies based on the organisation and the area being addressed.

The role of solution architect lies along two axes with four end points (see Figure 1.7):

- **Business:** business focus – for example, business analyst or business architect.
- **Technology:** focus on IT, automation and the physical aspects of software and hardware – for example, systems engineer or infrastructure architect.
- **Specification:** architectural responsibilities of responding to business needs and coming up with an appropriate set of solution requirements or a blueprint – for example, system designer.
- **Implementation:** delivering business change whilst balancing costs and resources, managing risk and, if necessary, trading off functionality with the delivery schedule – for example, project manager.

A closely related job role that shares these dimensions is the technical architect, which lies more towards the IT and implementation ends. The solution architect role lies closer to the business and specification end points.

Figure 1.7 Dimensions of the solution architect role

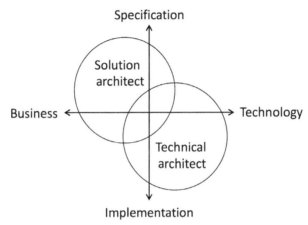

1.8 THE COMPONENTS OF A SOLUTION

Solving a business problem or taking advantage of an opportunity does not always have to be complicated or even require a complete solution architecture process.

Sandwich shop case study

- **Problem statement:** a takeaway sandwich shop has complaints that the hot products are not always hot.
- **Analysis:** observation reveals that this happens at busy times when the paninis, pasties and other hot items sometimes wait on the back counter too long.
- **Proposed solution:** the simple solution is for the kitchen staff to ring a bell when hot food is ready and for a server to hand it to the customer immediately.
- **Implementation details:** this required a few changes – installing the bell (technology), putting up a reminder notice (process) and talking to the staff (people).

There were several more complicated solutions available for the sandwich shop in the case study, but the sandwich shop owner chose this one because of the low cost and the perceived speed and simplicity of implementation (they did not use those exact words, but did have a bell at home that was not being used at the time).

Even this simple solution required changes to multiple areas of the business to be effective (technology, process and people). Two other areas where changes may be needed are information and organisation. The hot takeaway solution involved these two aspects but did not require any changes. The kitchen staff and servers were already in post (organisation) and they could already distinguish hot food from cold (information).

These aspects of a business or enterprise that may need to change are known as solution components. Changes include adding new components, modifying existing ones, or removing them completely. In order to achieve the change required, clear specifications of these component-based changes need to be documented so that the implementation and deployment can be governed successfully.

Resistance to change is a feature of everyday life, even when its benefits seem clear, and yet solving a problem requires change.

1.8.1 People

People play an essential role in any solution and without the motivation of staff members, customers and partners for a solution to succeed, many solutions fail to achieve the hopeful vision of their instigators.

- **Staff members:** people within the organisation who must change what they do, perhaps changing roles or routines.

- **Customers:** people who will use new or modified business services of the organisation or stop using defunct ones.

- **Partners:** outside agencies that must interact with the organisation to deliver new or changed business services or support changed processes.

- **Roles:** the parts that business actors play within the organisation may change by being redefined or new ones created from scratch.

- **Capabilities:** the things that people need to be good at to deliver business services and other components of the business model. Apart from new or modified (or deprecated) competencies, the solution may require an increased (or decreased) capacity or level of competency for capabilities involved in the solution.

- **Skills:** individual capabilities that may need to be developed in people, perhaps using a training programme if the scale or scope of change is large.

- **Job descriptions:** significant changes to the way people work need to be recognised by existing staff as well as being clearly communicated to new staff and those involved in people management. Job descriptions and other records are a formalised way of embedding and implementing changes in staff behaviour. These changes will endure when new appointments are made.

1.8.2 Organisation

The structure and behaviour of the organisation may need to change to enable the solution, particularly if it is part of a strategic move that has long-term implications.

- **Organisational units:** may consist of major divisions, functional departments and teams with specific responsibility, and the implementation and success of a solution may require a change (small or large) in the formation of these structures as well as the behaviour they produce. As with all solution components, organisational units may be created, modified or dispensed with.

- **Management structures:** reflect the control and communication of business activities and may follow certain patterns such as hierarchical, horizontal or matrix in order to coordinate how employees work together.

- **Outsourcing:** deciding which parts of the organisation are core and need to be performed directly by staff and which are peripheral is a strategic decision that can have an enormous and wide-ranging impact. The reverse of outsourcing is insourcing. Almost all solutions are a combination of multiple components. A single change, such as outsourcing or insourcing, is unlikely to solve a problem or capture an opportunity. A holistic approach to change is usually required.

- **Partner organisations:** includes organisations performing outsourcing activities but also suppliers, intermediaries and government agencies such as regulators. These are all required to achieve the mission and vision of the organisation. Solutions very often have an impact on partner organisations that must be taken into account in the solution.

- **Targets and service-level agreements (SLAs):** these are the measures and contracts that allow managers to decide if a solution is working and to take action if not. A solution may not work as planned without requiring a change in the level of performance by part of the organisation. Therefore, new or modified targets and SLAs are often included as components in the architecture of a solution.

1.8.3 Processes and procedures

Potentially cutting across multiple areas of the organisation and involving a variety of people, processes and procedures embody the activities that are carried out to achieve the mission of the enterprise.

- **Business service:** is what the customer wants and hopefully receives from the organisation. There is sometimes a distinction made between product and service, but this has blurred in recent years so that the delivery or customisation of a product is seen as a service and some organisations call their services 'products'. This is the point of interaction with customers.

- **Value chain:** is the set of activities that an organisation (or organisational unit) carries out to create value for its customers (Porter, 1985). This is an end-to-end process that may be broken down into smaller business processes. Some solutions require the creation of an entirely new value chain or specify modifications to an existing one.

- **Business process:** is a set of linked activities and tasks that together achieve a business objective or goal. Business processes may form part of a value chain. They may directly or indirectly support and enable the delivery of one or more business services.

- **Policy:** guidelines developed by an organisation to govern the actions of staff and business units, such as defining the limits within which decisions can be made and often including controls to avoid issues. Policies often apply across multiple business processes and services and affect staff in multiple organisational units as well as external organisations and people. Policy changes and sometimes entirely new policies are often used to reinforce behavioural changes and may be linked to external regulation and legislation.

- **Business rule:** a condition that controls whether a business activity is allowed and, if so, how it must be performed. They are often part of a policy.

- **Procedure:** a specified way to carry out an activity or task. Overlaps with business process, policy and business rule and may cover multiple processes. Procedure is often too coarse-grained to be a component of a solution but may be listed as a dependency.

1.8.4 Information and data

In the digital world, information and data are constantly increasing in volume, velocity and variety and, for businesses, digital forms of communication are largely taking over from traditional person-to-person relationships with customers. Organisations often state the desire to be 'data-driven'. Automated data processing by algorithms and artificial intelligences (AIs) is the order of the day. Tasks such as data analytics and visualisation are being performed by practitioners of the relatively recent discipline of data science.

With this dominant position in the business landscape, it would be surprising for any strategic change to be considered without an analysis of the information and data aspects.

- **Input:** information and data that is collected from a person or system as part of a business process.

- **Output:** records and business intelligence (BI) that result from the operation of the solution and that are valuable to others in the organisation or outside, such as customers or business partners. Records are usually stored for future use by the same or another solution.

- **Trigger:** information or data that, either by its arrival or a change in state, stimulates an action by a human or system actor.

- **Feedback:** something that assists business actors in performing a task or business process.

- **Audit:** a type of record, including business transactions, decisions and changes to data, that can be used for compliance with internal policies and external regulation.

- **Information as a product or service:** a type of output that is valuable to customers and that it may be possible to monetise.

1.8.5 Technology

Technological innovation permeates the modern world and affects many areas of business. Technology may be an end in itself, if it is created and sold, but for every organisation it is an enabler for the vision and strategy that the business wishes to pursue.

It is unusual (but not impossible) for a solution not to have within it some technological components in one form or another. Note that technology is not limited to the use of computers (although they are included in this category) but extends to nearly every sphere in which businesses operate.

- **Manufacturing, storage and transport:** are all areas where robotic hardware and software and control systems have enabled automation in processing and quality checking. Many components for solutions in these areas of industry are now available.

- **Tools and materials:** the availability of new materials and improvement in existing ones mean that solutions that were previously impossible or unaffordable are now in view for even small and medium-sized organisations. Similarly innovations in tools enable the same work to be done in less time, to a higher standard or with less training.

- **Computer hardware, software and communications:** continue to provide improved services in terms of functionality, reliability and reach.

- **Internet of things (IoT), cloud and big data:** have all provided innovations that make new solutions possible. In many cases these have provided a technology-driven stimulus for solution requirements.

Activity 1.4

Identify one component that is likely to be part of the Fallowdale Hospital patient communication solution from each of the five categories (people, organisation, process, information and technology).

1.9 OBJECTIVES AND BENEFITS OF SOLUTION ARCHITECTURE

Some of the key objectives and benefits of using an architectural approach to designing and implementing a solution include solving the problem with a holistic approach, strategic alignment, efficiency, and embracing the progressive nature of change. These are all areas that enhance organisational development and add value to the business as a whole.

1.9.1 Holistic approach

The idea of a solution, at least for any non-trivial problem, implies a multi-part approach involving business and IT working together. This can be expressed using the acronym POPIT, standing for People, Organisation, Processes, Information and Technology (Cadle and Paul, 2021). Very few successful solutions are deployed without some change in each of these areas.

Solution architecture treats the solution as a system that functions to solve the problem. Note that by thinking more positively, solutions may also enable an organisation to take advantage of opportunities.

As with the architecture of any system, it has some behavioural aspects and a structure.

Solution architecture looks at the system holistically and not only accepts that it contains elements from POPIT but also aims to show how these elements can work together efficiently and effectively. A proposed solution architecture therefore may include building blocks from any part of the business, including customers and partners, as well as more technical components such as data, information and technology.

1.9.2 Strategic alignment

The motivation for a solution may originate from the organisation's strategy or be linked closely to it, especially if the original idea behind it comes from a formal framework such as enterprise architecture or business analysis. If not, the solution architecture process looks for strategic direction in all its design decisions and avoids anything that makes changes that conflict with the organisation's strategy.

The discovery phase of the solution architecture life cycle identifies business requirements that are strategic and takes a long-term view of the way the solution will work to achieve those aims.

Any short-term, urgent solutions that go against strategy can be identified as **strategic debt** and plans can be made to 'repay the debt' or strategically realign the solution as soon as possible.

1.9.3 Reuse of resources

Solution architecture is based on the use of component-interface models, which increases the likelihood of existing components being identified as suitable for multiple purposes and therefore reused.

1.9.4 Evidence-based decision making

The architectural approach models the problem area by analysis that enables the root cause to be located and ideally isolated, so that it can be addressed directly. This systematic approach is based on evidence rather than opinion and therefore has a better chance of success.

Proposed solutions are also modelled to test whether and to what degree they will solve the problem and, separately, whether they address the concerns of stakeholders. Any variations based on competing concerns can be assessed according to the evidence from these models.

1.9.5 Minimal cost and disruption

The systematic and structural approach to the analysis and design of solution architecture means that multiple potential solutions can be compared using **gap analysis** to find the smallest change that will achieve the desired outcome.

Change can rarely be achieved with zero disruption, but it should be limited as far as possible. Again, the systematic approach of solution architecture provides a way to do this by using **impact analysis** to limit the side effects, enabled by the structural models that show all areas affected by the change. Two or more proposed solutions can be compared by the scale of their impact and this could be used as a factor in deciding between them.

1.9.6 Reduced dependency through encapsulation of components

Dependency is the concept that a change in one area of a solution necessitates a change somewhere else. This can add significantly to the cost of change either now or in the future. One response to this is to design self-contained components with defined interfaces. Each component represents an area of the solution. If the interfaces are maintained without change then any internal aspect of that area can change without any knock-on effect. This is known as encapsulation of components.

1.9.7 Early resolution of conflict and duplication of effort

Areas of potential conflict are identified throughout the solution architecture process by a number of techniques. The scope of the solution is defined in terms of its architecture and the areas that will change are identified by gap analysis. Other areas of the business that will be affected are identified by impact analysis. This makes it more likely that any other solutions, programmes or projects that are proposed or ongoing in these areas can be highlighted for investigation to see if any conflict or duplicated effort is likely to arise as a result of the solution.

In more general terms, the clarity of communication and the involvement of stakeholders in the design of a solution mean that conflicts and duplication of effort will be spotted early, and informed decisions made about how to proceed.

REVIEW QUESTIONS

1. What is the name of a work product used to express an architecture?
   a. Architecture documentation.
   b. Architecture description.
   c. Enterprise architecture.
   d. Domain architecture.

2. Which **three** levels may be used to break an organisation down into successively smaller, nested systems?
   i. Enterprise system.
   ii. Business system.
   iii. Organisation system.
   iv. Information system.
   v. Information technology system.

   a. i, ii and iii only.
   b. i, iii and v only.
   c. ii, iv and v only.
   d. iii, iv and v only.

3. Why is solution architecture described as holistic?

   a. It attempts to solve all an organisation's problems at the same time.
   b. It is unlimited in scope.
   c. It covers multiple aspects of the business, not just IT.
   d. It does not distinguish one solution from another.

4. Which artefact is the main handover between the design work of solution architecture and the implementation work of programme and project management?

   a. Delivery roadmap.
   b. Implementation plan.
   c. Solution itinerary.
   d. Project initiation document.

2 SOLUTION ARCHITECTURE IN THE CONTEXT OF BUSINESS AND ENTERPRISE ARCHITECTURE

LEARNING OUTCOMES

When you have completed this chapter, you should be able to demonstrate an understanding of the following:

- How enterprise architecture relates to solution architecture
- Granularity of architecture
- Business architecture
- Business drivers for solution architecture
- Business components of a solution
- Enterprise and solution data architecture
- Generalisation and specialisation of data and information
- Applications architecture
- Infrastructure architecture
- Software architecture
- Security architecture

2.1 HOW ENTERPRISE ARCHITECTURE RELATES TO SOLUTION ARCHITECTURE

Enterprise architecture comprises the combined business, IT and other architectures and processes that relate to an **enterprise**, which could be:

- An **organisation** such as a corporation, start-up, government department, charity or other category.
- A **division of a large organisation**, such as one arm of a multinational company or a geographical region that has a degree of autonomy.
- A **collection of organisations** working together, such as an industry group or intergovernmental consortium.

Solution architecture can be used to address business problems in the context of these enterprise types and sizes. Therefore a solution is usually built to address the needs of an enterprise.

Enterprise architecture is the highest-level, most strategic type of architecture and its directives, such as **principles**, **policies** and **business rules**, must apply everywhere within the enterprise, including within solution architecture and the resulting solutions.

There may be some, often tactical, solutions that do not follow the directives of enterprise architecture. Good governance demands that these exceptions are only allowed when there is no alternative, that measures are taken to mitigate any negative effects, and that the directives are incorporated in the solution, or its replacement, as soon as possible. These temporary exceptions go against the strategic decisions that put the directives in place, so this approach is considered a strategic debt that needs to be corrected or eliminated in the future.

Fallowdale Hospital: core principles

- Providing services fairly and equally, based on clinical need.
- Demonstrating excellence and professionalism.
- Putting the patient at the heart of everything.
- Working across organisational boundaries.
- Providing best value for taxpayers' money.
- Being accountable to the public and our community.

These principles must apply to the entire organisation, including any solutions being designed.

Enterprise architecture may be divided into four subdomains:

- **Business architecture:** everything to do with the business, including products and services, business processes, organisational structure etc.
- **Data or information architecture:** everything about how the enterprise uses information and data to deliver its business operations.
- **Applications architecture:** how software applications support business operations and help to manage information and data.
- **Infrastructure or technology architecture:** details and models of all infrastructure and technology that underpins the operation of software applications, data management and any other aspect of business operations.

Each of these subdomains has specialist **domain architect** roles with responsibility for developing and maintaining the architecture. Depending on the size of the enterprise, some of these roles may be combined.

Solution architecture uses components from all of these domains due to its holistic approach to designing complete solutions that work with and connect to all aspects of the business. For example, a solution design may include business processes, application software, data structures and infrastructure services.

There are two other widely recognised specialist architecture domains that exist within the scope of enterprise architecture:

- **Security architecture:** closely linked to infrastructure architecture because of the risks around cyberattacks, but also provides security directives that are enterprise-wide and apply to all four subdomains: business, data, applications and infrastructure.

- **Software architecture:** concerned with the design of software components, subsystems, interfaces and applications, each of which may be included as part of infrastructure, applications and solution architecture.

One of the objectives of enterprise architecture is steering the organisation in the strategic direction of its mission. This can be achieved by identifying problem areas and commissioning a solution architect to design a solution that is perfectly optimised for the enterprise. In this sense, enterprise architecture is the governing authority that directs all architectural activities, including solution architecture (see Figure 2.1).

Figure 2.1 Solution architecture in the context of other architectures

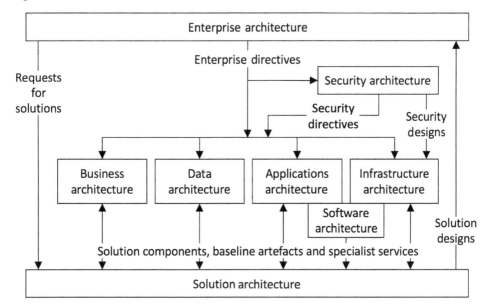

Enterprise architecture also aims to ensure consistency and completeness of understanding in order to aid decision making. This global view could affect the scope of the solution.

Fallowdale Hospital: scope of solution

In the Fallowdale Hospital patient communication scenario, the initial imperative is to solve the problem of outpatient appointment letters. Following discussions with enterprise architecture, a decision has been made to broaden this out to cover other areas of communication, such as reminders and notices to patients with chronic (long-term) conditions and more general information about hospital services.

Enterprise architecture models and artefacts cover the whole organisation and can therefore be used to understand the wider impact of making changes, including implementing new solutions. This supports the aim of solution architecture to minimise the negative impact of change.

Perhaps the biggest business impact of the practice of enterprise architecture within an organisation is that it acts as a source of knowledge that enables problems to be diagnosed and solutions designed using solution architecture. Many invaluable artefacts will be available in the enterprise repository, which can act as a source of information for:

- business requirements;
- relevant architectural models;
- stakeholders;
- previous solution designs;
- building blocks and interfaces;
- viewpoints and views;
- impact models;
- gap analysis models.

The enterprise architecture repository is the ultimate storage and reference location for all solution architecture artefacts once the solution has been signed off. Note that this is also known as the enterprise continuum.

The Open Group distinguishes between the enterprise continuum, which 'categorises architectural source material' and the architecture repository, which is 'The practical implementation of the Enterprise Continuum' and 'includes reference architectures, models, and patterns that have been accepted for use within the enterprise, and actual architectural work done previously within the enterprise.'

(TOGAF 9.2, 2018)

A number of enterprise architecture frameworks are available:

- **DoDAF:** United States Department of Defense Architecture Framework, optimised for military and defence concerns (DoDAF 2.02, 2010).
- **Zachman:** a generic framework that is easily adapted to any organisation as it does not have any specific content (Zachman, 2008).
- **TOGAF:** The Open Group Architecture Framework (TOGAF 9.2, 2018).

Other specialised enterprise architecture frameworks are available and may be used to support solution architecture.

Activity 2.1

Identify one or more enterprise principles or policies of Fallowdale Hospital that could apply to the patient communications solution and explain your reasoning.

2.2 GRANULARITY OF ARCHITECTURE

Architecture operates at multiple levels of abstraction, with enterprise architecture at the highest level. Enterprise architecture directives such as policies and principles apply throughout the enterprise and must therefore be highly generalised. More specific designs, standards, models and other architectural artefacts are organised into the subdomain architectures of business, applications, data, infrastructure and security.

Subdomain architectures apply across the entire enterprise even though they are more specific about what they define. For example, data architecture may define all aspects of the data structures associated with the customer entity, whereas the business architecture defines the business services, processes and other parts of the business that involve customers.

Bank case study

A bank has two enterprise architecture principles: that the customer is at the centre of business operations, and that there are to be no boundaries between different parts of the business. The data architecture reflects these principles in its design of a single view of the customer, with all aspects of data linked to a single customer entity.

At the more specific level of solution architecture, certain aspects of the domain architectures may be refined, and more detail added that is only relevant to the solution.

The bank has commissioned a new solution to facilitate sharing resources within a household or family unit. For consistency, and to preserve adherence to the enterprise architecture principles, the starting point for anything to do with customer data in the solution design must be the existing data architecture for customers. New, more specific data items may be added to the solution data architecture if required and must be considered for inclusion in the enterprise data architecture based on relevance to other areas of the business.

An even more specific level is software architecture. This is concerned with the design of software components that form part of a solution or application, or provide an infrastructure service.

The bank has specified a number of software components that will provide functionality for the resource-sharing solution. Each software component must base any data handling on the solution data architecture but will add specific details concerning data storage, manipulation and messaging, whilst complying with any directives from other architectures, such as security, applications and infrastructure.

This approach ensures consistency between architectures and compliance with directives at every level.

Architectural granularity is a conceptual model that represents the different levels of architecture within an enterprise. In this model, an enterprise is composed of a number of solutions that address specific areas of the business. Each solution is composed of a number of components that may belong to a subdomain architecture or may be a software component (see Figure 2.2).

Figure 2.2 Architecture granularity

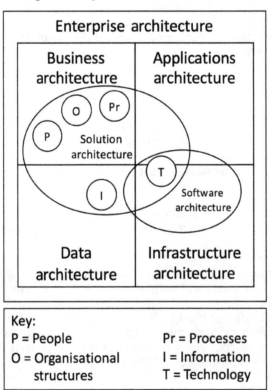

Solution architecture is holistic because it covers all aspects of the solution to a business problem, not just IT systems. Solution architecture accounts for the fact that a solution depends on people, organisational structures, processes, information and IT, known as POPIT (Cadle and Paul, 2021).

Solution architecture is predominantly focused on satisfying business requirements and specifying how these can be achieved at a logical level. This means that solution architecture overlaps with three of the enterprise architecture domains: business, data and applications.

Inevitably, part of a solution in the digital age will comprise one or more software components and there is an architectural discipline concerned with these, called **software architecture**.

Software architecture aims to deliver the requirements of a solution at the physical level and overlaps with:

- **Applications architecture:** the human and system interfaces for software.
- **Data architecture:** the design of data and information structures that are managed by the software.
- **Infrastructure architecture:** system software and hardware that enables the software to function and communicate.

Other types of component fall mainly within other architectural domains, although all components are connected, so there is some overlap:

- **Business architecture:** people and roles, organisational units and business processes.
- **Data architecture:** information and data.

Fallowdale Hospital: existing solutions

Fallowdale Hospital is seeking an improved solution for patient communications but already has many solutions, such as one for rostering nurses on to inpatient ward shifts, 24/7, at safe staff-to-patient ratios.

2.3 BUSINESS ARCHITECTURE

Business architecture: business architecture as 'A blueprint of the enterprise that provides a common understanding of the organization and is used to align strategic objectives and tactical demands'.

(Business Architecture Guild, 2021)

The TOGAF definition is 'A representation of holistic, multi-dimensional business views of: capabilities, end-to-end value delivery, information, and organizational structure; and the relationships among these business views and strategies, products, policies, initiatives, and stakeholders' (TOGAF 9.2, 2018).

Business architecture is about analysing the business and making changes to bring about a targeted and sustainable improvement. One way of achieving this is through restructuring the business directly or by asking another domain architect to make changes, for example, to improve the consistency of data or the performance of an application.

Another, perhaps better way is to use solution architecture to deliver a more comprehensive, all-encompassing solution to the business problem, risk or opportunity that has been identified.

When to use solution architecture

Solution architecture is an excellent approach for many situations in business. However, some extremely complex problems may not suit its analytical approach, whereas others are too simple to justify such a systematic method.

Solution architecture relies on the idea of modelling a system and reducing it to its component parts. To address a problem using this technique, it must be possible to produce a model and use it to show that the behaviour of the system can be changed to affect the problem. The model is the basis for all the techniques used in solution architecture, including finding the root cause of the problem and isolating the components that can be altered to change the behaviour of the system.

When deciding whether solution architecture can be used to solve a problem, the following questions must be considered:

- Is the problem area a system?
- Can the system be modelled?
- Does the model exhibit the problem?
- Can the model be altered to address the problem?

If the answer to any of these questions is 'no', then solution architecture is unlikely to be the best way to address the problem.

The problem area may be too complex for solution architecture and need an alternative approach, but it may just be too big. It is worth trying to reduce the scope and solve a smaller problem. For example, perhaps the problem can be addressed for a limited range of products, a single geographical zone, or a contained group of people or organisational units.

There are many alternatives to the design-based approach of solution architecture. Rather than trying to find the cause and potential solution to the problem using a model, a common method is to make a change based on the suspicions and intuition of those involved. By waiting a sufficient time after making the change, it can then be judged whether the problem has been eliminated or reduced. This approach is more appropriate for complex situations that may be a system but cannot be reduced to a model that replicates the unwanted behaviour.

Sometimes the problem area is too simple to be called a system, or to require modelling. In these cases, solution architecture is likely to add complexity rather than helping to reveal the cause of the problem and should not be used.

In both cases of high complexity or extreme simplicity, some of the techniques of solution architecture can still be useful but the full life cycle should not be followed.

2.3.1 Business architecture models

Some key models that are used to describe the business architecture of an organisation are:

- **Business model:** this can take many forms and can include many artefacts, although a popular one is the business model canvas (Osterwalder et al., 2010). Whatever form the business model takes, it aims to give an overall picture of the business to aid comprehension between stakeholders and decision makers within the organisation.

- **Business motivation model**: links components of a strategy or business plan together logically to identify the motivation for change (Formal/2015-05-19, 2015).

  The starting point is called **influencer**, which is also known as a driver for change. Influencers can be internal, identified by the business as a problem or opportunity for improvement, or external, representing a change in the business's operating environment.

  Each influencer is given an **assessment** of its potential impact on the business. This assessment feeds into both the **ends,** which represent the objectives of change, and the **means**, or methods of achieving the objectives.

  The model does not dictate any method of progressing the plans but does stimulate analysis and acts as a record of the organisation's thinking about them.

  Once the organisation has decided that change is required in a certain area of the business, one way of bringing it about is through solution architecture.

- **Capability map:** a capability is what an organisation must be able to do to deliver business services and execute its business strategy. Note that it may not exist at present but be part of a capability plan. Capabilities have additional requirements for a level of capacity (how much can be delivered in parallel) and competency (how skilled the organisation needs to be at doing this). Each capability is enabled by one or more resources, including people, organisational units, technology and specialist knowledge, which could become components of a solution. Capabilities enable or support business activities such as business processes and services which can also be solution components.

  The baseline artefact for capability-based planning is a capability map that breaks down top-level capabilities into smaller ones and classifies them as strategic, operational or supporting. Many other classifications can be used including maturity and core versus non-core.

If a new capability is required, a capability map can be used to provide a model showing the resources (components) required and activities supported by the capability. This approach will give a complete picture of the change, including other areas of the business that could be affected. Some of the required resources may already exist elsewhere in the organisation, for example.

The need for changes to capabilities can be driven from strategy, for example the use of a business motivation model.

- **Value stream map:** an end-to-end set of actions that provides value to a customer, from initial interaction with the business to realisation of the value by the customer. Organisations may have multiple value streams. The technique of mapping value streams comes from the Lean methodology (Ohno, 1988) which seeks to eliminate waste, that is, any steps that do not add sufficient value. The Lean approach is often combined with Six Sigma (Harry, 1988). Six Sigma also looks at each step and asks whether a higher-quality output can be achieved.

  Value stream mapping (VSM), and the analysis of the steps involved, may reveal the need to modify or remove some of the steps. Solution architecture works well with VSM because they both analyse a problem in terms of structure and behaviour, and it is therefore straightforward to take the requirement for change identified by VSM and design a holistic solution using solution architecture.

- **Functional decomposition:** the structure of an organisation based on the functions performed. An organisation chart is an example, providing it shows what each unit does. It can be cross-referenced with value stream and capability maps to check alignment with strategic planning.

- **Business process models:** a typical building block of solutions, process models show business activities broken down into steps and who is responsible for performing them.

  Please note that this is not an exclusive list but represents the most frequently used models. Analysis of any of these models may lead to ideas for improvement that could be delivered through solution architecture.

Fallowdale Hospital: capability-based planning

Fallowdale Hospital currently has the capability to communicate with patients but only in a limited way. Its plan is to increase the capacity so that more communication can take place and to add flexibility so that communication can use multiple channels and be more timely – for example, to remind patients the day before an appointment.

2.3.2 Life cycle for business change

Many business analysts and architects use a five-stage life cycle for business change (BCS, 2019). The five stages are known as align, define, design, implement and realise (see Figure 2.3). The use of this life cycle can lead to ideas for solutions that can be taken forward by solution architecture.

Figure 2.3 The life cycle for business change

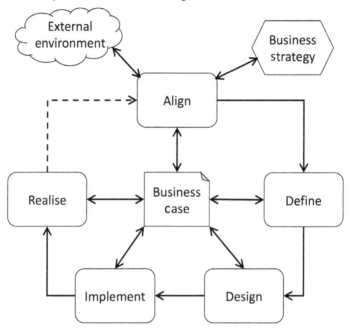

- **Align:** examines the external and internal business environments for opportunities to align with the external environment and its internal processes and systems, a potential source of solution concepts.

- **Define:** identifies business requirements and constraints for potential solutions, as well as articulating the benefits and creating a **business case**.

- **Design:** the main focus of solution architecture.

- **Implement:** where solution architecture interfaces with P3M to produce the solution, providing the delivery roadmap and retaining a governance role.

- **Realise:** the point of handover of the solution to the business, during which solution architecture continues in a governance capacity to ensure a satisfactory delivery.

Solution architecture is mainly concerned with the define and design stages of this life cycle.

2.4 BUSINESS COMPONENTS OF A SOLUTION

Business architecture provides many artefacts as inputs to solution architecture and a number of techniques through which the need for change in the business domain can be identified. Some of these techniques use similar models, based on similar components to solution architecture, and therefore work well as a source of solution ideas.

The business architecture domain is also the target for the change brought about through solution architecture. Three principal categories of components that form part of the solution design are also part of business architecture: people, organisation and business processes.

2.4.1 People

Some solutions are entirely automated but many involve people, such as:

- staff within the enterprise;
- customers;
- people at third-party organisations, such as business partners.

Some aspects of people that can be defined as part of a solution are:

- roles performed;
- job specifications;
- skills and capabilities.

Fallowdale Hospital: solution components involving people

People components include:

- Patients and their communication preferences and habits and what will work for them.
- Clinic administrators and healthcare professionals who need to be confident in the communication process and kept informed by the solution to suit their way of working.
- People who will decide what is communicated and when.

2.4.2 Organisational units and hierarchy

A major component of a solution is the organisational unit, which could be a team or department that performs tasks or interacts in some other way with the other components. Organisational units can be within the organisation or external to it, such as business partners or outsourced functions. An important aspect is the management of such units by a management structure and by setting targets or through service level agreements (SLAs).

Fallowdale Hospital: organisational solution components

Organisation components include:

- Post room, which owns the letter printer and enveloping equipment and will need to reduce its capacity to handle letters and to use the equipment for other purposes.
- Outpatients department and individual clinics, which currently dedicate large numbers of staff to dealing with cancellations, rebooking and last-minute requests.

2.4.3 Business processes

Business processes are perhaps the best-known artefact from business architecture to be a component of a solution.

Processes can be broken down into activities that are usually defined as something that a person or machine does as a unit of work, continuously and in a relatively short time period. These activities and the steps involved may be the subject of improvement in terms of efficiency or automation as part of the design of a solution.

Processes are governed by directives such as principles, policies, procedures and business rules that may have been created specifically for the process or apply more generally across the enterprise.

All solution components, including business processes, ultimately support one or more business services.

Fallowdale Hospital: business process solution components

Process components include:

- Appointment booking.
- Cancellation and rescheduling of appointments.
- Cancellation and rescheduling of clinics.
- Waiting list management.
- Management information reporting.

Activity 2.2

A supermarket has identified that environmentalism is important to its customers. According to research, customers particularly dislike non-recyclable packaging. Goods are sourced from multiple suppliers under contract. Warehousing and distribution are performed by the supermarket. Some goods are repackaged by the supermarket.

The value chain of selling goods to customers comprises the following steps:

- product selection;
- supplier selection and contract agreement;
- call-off of supply;
- supply of goods;
- warehousing;
- packaging of unbranded goods by supermarket;
- distribution to store;
- purchase by customer.

Which of these steps are likely to need changing in order to address the issue of recyclable packaging?

2.5 ENTERPRISE AND SOLUTION DATA ARCHITECTURE

Data architecture is a subdomain of enterprise architecture and is concerned with the data, metadata and information used within the enterprise. An important high-level artefact is an enterprise (or corporate) data model. This provides the 'big picture' of the use of data and information within the organisation and helps stakeholders from diverse areas of the business to understand what is available and how it is being used.

Data and information

These two terms are often used interchangeably but they can be distinguished. The following definitions are given by ISO IT vocabulary (ISO/IEC 2382:2015, 2015):

- **Data:** reinterpretable representation of information in a formalized manner suitable for communication, interpretation, or processing.
- **Information:** knowledge concerning objects, such as facts, events, things, processes, or ideas, including concepts, that within a certain context has a particular meaning.

Information then has meaning within a context, whereas data is a representation of information that is suitable for processing and communication. Data can be converted to information through interpretation. Conversely, information can be represented using data that can then be reinterpreted to produce different information.

Another way of looking at information is that it is unstructured. Examples of this are video, audio and text documents. These clearly contain meaning but are not structured as data and so are not easily processed or reinterpreted, except with human involvement. Artificial intelligence (AI) can convert this type of information to data with techniques such as voice and visual recognition combined with interpretation, often based on machine learning.

On a more philosophical level, information can be raised to higher levels such as knowledge, intelligence or even wisdom, by further interpretation.

Metadata is an intermediate category that is commonly defined as 'data about data' and this provides a formalised means of interpreting the data. Metadata includes descriptions, rules, constraints and links between different data items or entities.

The terms 'data architecture' and 'information architecture' are both in use and most organisations choose to use one or the other. Note that this approach does not deny the difference between data and information but combines them into a single architecture because they are so closely related. This book uses data architecture to include information and other business uses of the data.

Solutions have a data architecture that should be consistent with that of the enterprise in which the solution will operate. If there is inconsistency, either in the design and definition of the data or its use and interpretation, there is a risk of error in a business service or of missing an opportunity that would benefit the organisation.

Video production case study

A video production company has details of the authors of video content who are either staff members or external contractors. The sales and marketing department has details of customers and products.

Some external contractors are also customers of the company, but the definitions of **customer** and **author** are different and inconsistent with each other. When a contractor contacts the company to notify a change of contact details, this is recorded for either the person as a customer or the person as an author, but not both.

As a result, the company may try to contact the person using incorrect details.

Consistency between solution and enterprise data architecture requires that any new solution has any relevant components of the enterprise data architecture as inputs. This reduces the risk that existing data definitions and structures will be recreated from scratch, thereby introducing inconsistency.

New solutions often require new data that is not in the current architecture, so data designs that are part of solution architecture need to be integrated into the enterprise data architecture as soon as possible, even if in a provisional state, in order that other architects can be aware of the new data.

2.5.1 Objectives

Data architecture has several objectives which overlap with or support solution architecture:

- Meeting the data and information requirements of the business, both strategic and operational.
- Promoting understanding of data throughout the enterprise.
- Maximising the effectiveness and efficiency of data use through continuous improvement.
- Ensuring consistency of data use and eliminating duplication of data definition.
- Complying with data legislation and regulations.
- Governing the development of new solutions.

2.5.2 Activities

Many of the key activities of enterprise data architecture are concerned with continual improvement through rationalisation and redesign to support the aims and objectives of the business. Others are more directly related to solution architecture, including:

- Recording and integrating data definitions and models.
- Supporting the design of new or revised data definitions and models.
- Auditing data designs and architecture against standards, including legislation and regulations.
- Providing views (specifically constructed subsets) of the data architecture to assist in solution design, data analysis and other data-related tasks.

2.5.3 Artefacts

Enterprise data architecture artefacts are similar to those of data analysis – for example, entity relationship diagrams (ERDs) – but tend to be at a higher level of abstraction. Solution data architecture is naturally more specific and focused.

Data entities and structures are components in a solution and so existing enterprise data architecture artefacts that contain relevant entities are essential inputs to solution architecture:

- models and schemas of data, including messages and streams;
- data definitions;

- data dictionaries and catalogues;
- data store designs for databases and data analytics.

Some artefacts are developed as part of solution architecture, such as data requirements, and can be checked and validated by data architecture.

There are also more general artefacts that are refinements from enterprise architecture, including:

- principles and policies related to data;
- standards, legislation and regulations.

A category of artefact that is particularly useful for impact analysis is the matrix. These are used to cross-reference data entities with various other components such as business functions or software applications. During impact analysis, it is easy to see which other components will be potentially impacted by a change to a data entity.

2.5.4 Relationship to solution architecture

A complete data architecture is required as part of the design of a solution covering the domain of the solution. The elements to construct this data architecture may exist in the enterprise or corporate data architecture or may need to be created. The data architecture function will be able to assist with both of these activities.

In addition, the data architecture function can provide details of any differences between the existing data architecture and any changes that will be made as a result of the proposed solution design. These differences will be a key input to the gap analysis activities that will be performed during solution architecture.

The data architecture function will also be able to provide an analysis of the impact of any changes to the data architecture and this will be a key input to an overall impact analysis for the proposed solution.

The data architecture function may also be able to assist with providing or designing viewpoints that can be used to produce views of the proposed solution that address specific concerns of stakeholders.

As data is a major resource that needs to be protected, the data architecture function provides vital information that informs the security design and architecture of a solution.

2.5.5 Generalisation and specialisation of data and information

Data and information are a type of abstraction of things in the real world. The information held on record for processing, about a customer for example, only contains details that the business is interested in rather than being a complete representation of the person. Legislation on personal data constrains this further by requiring there to be a business need to hold or process data and also that the person has given consent for their data to be used. Examples include the UK's Data Protection Act (Data Protection Act, 2018) and the General Data Protection Regulations (GDPR) of the European Union (Regulation (EU) 2016/679, 2016).

The first type of abstraction is to model a real-world thing or entity as a data entity.

Fallowdale Hospital: data entities

The Fallowdale Hospital patient communication solution sends invitations to patients to attend clinic appointments with a healthcare professional. From this description, a data architect would extract **patient**, **invitation**, **clinic**, **appointment** and **healthcare professional** as concepts that need to be modelled as data entities.

The second type of abstraction is to generalise several distinct data entities to a single, more abstract data entity.

Fallowdale Hospital: generalisation

Patient communication involves two entities – **patient** and **healthcare professional** – that have some similarities and these could be generalised to a **person** entity.

Both of these techniques help to match data in the problem or solution space with existing data definitions and models in the data architecture. Note that the data entities may not have the same name, so thinking in abstract generalisations can help to see that there is a data or information component that can be reused.

Data architecture is also concerned with making data sufficiently generalised so it is appropriate for multiple solutions and other uses.

The reverse of generalisation is specialisation. This means taking a data entity and modelling it as a more specific entity. This is appropriate if there are genuine differences in the behaviour or structure of the new specialised entity.

Fallowdale Hospital: specialisation

The **healthcare professional** entity has been identified, but there are significant differences in the behaviour and structure between **specialist nurse** and **doctor**. Therefore two new specialised entities have been modelled. Note that the new entities are still classed as healthcare professional.

Activity 2.3

An organic fruit farm is looking at ways to improve yields through monitoring its plants and their growing conditions so that it can regulate food, water and other environmental factors.

Identify some data entities that fall within the problem or solution space.

2.6 APPLICATIONS ARCHITECTURE

Applications architecture provides a portfolio view of the enterprise's IT applications and the application services provided by them. An **application** is a set of technological (software and hardware) capabilities that provide key business functions and manage data assets. An **application component** encapsulates a unit of functionality and provides its services through clearly defined interfaces.

Applications architecture is a subdomain of enterprise architecture that links the business and data architecture domains. Applications support business processes through partial or complete automation, for example by providing information to enable a person to make a decision. Applications are the primary method for enabling the management of data through its life cycle of acquisition, processing, storage, presentation, archival and deletion.

These relationships can be arranged in a hierarchy of layers, each of which interfaces with the immediately adjacent layers (see Figure 2.4). The top layer, business services, is the interface with the customer and is enabled through one or more business processes. Application services provide automation support for business services using a number of application components. Application components in turn rely on technology services such as storage, networking and processing; these are provided by technology components such as servers and system software.

Figure 2.4 Architecture hierarchy

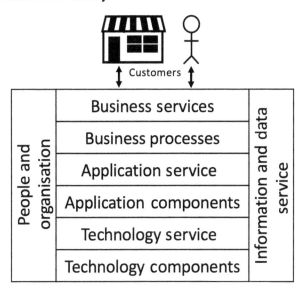

This hierarchical organisation of services and components into layers allows the separation of functionality, so that any service or component is only dependent on one layer below and only depended on by the layer above. This structure is modular and replaceable and encapsulates the behaviour of each item through its interfaces. It also promotes reuse, as each component or service may support multiple items in the layer above.

2.6.1 Objectives

The primary objective of applications architecture is to ensure that the organisation has the necessary set of applications available to support the business operations and processes. The set of applications is called the portfolio and is catalogued and described in detail in the primary artefact of applications architecture, the applications portfolio catalogue.

The applications portfolio is usually a complex collection of applications that have been acquired or built at various times in the organisation's history, including through internal restructuring and mergers with other organisations. Therefore, application architecture has a constant remit to rationalise and reorganise the portfolio to better serve the business needs, both functional and non-functional.

The applications portfolio also needs to support strategic change through flexibility and the selection and use of the ideal set of applications for a given situation. Application architecture can distinguish between strategic and legacy applications and often uses a categorisation such as red–amber–green to label applications. This allows solution architecture to select the most strategic option.

2.6.2 Artefacts

The primary applications architecture artefact is the **applications portfolio catalogue**, which is a list of all the applications in the enterprise. This gives an overview of the applications in use and planned within the organisation with details of how and by whom they are being used.

A closely related artefact is the **applications interface catalogue**, which documents all the interfaces between applications. This shows the level and type of dependency between applications and gives details of how they interact, including their application programming interfaces (APIs). This may be modelled using an **applications communication diagram**.

More detail of the functionality of each application is documented and modelled with **use case** and other **functionality models** and **realisation diagrams**.

Applications architecture has artefacts that represent the relationship with business and data architectures in the form of cross-reference grids:

- data to application;
- business function to application;
- business process to application.

2.6.3 Relationship to solution architecture

Applications are likely to form part of a solution, working closely with business and data architecture components. If part of the solution functionality can be provided by one or more existing applications, and provided these are part of the future applications architecture strategy, then they should be considered as building blocks of the new solution.

Examination of other artefacts, such as the interface catalogue and communication diagrams, will highlight potential integration requirements for the new solution, as well as identifying the wider impact of making changes. The applications interface catalogue is a key source of information for interface analysis during the solution architecture life cycle. Using application architecture cross-reference grid artefacts can also help with impact analysis.

Functionality that is required by the solution but not provided from within the existing applications portfolio may be used to identify the need for new applications or modifications to existing ones and may have a knock-on effect on business processes and data components.

Types of application that may be required include:

- **Business application**: directly linked to business activity; may be unique to the organisation or off the shelf, such as enterprise resource planning (ERP) or customer relationship management (CRM) systems. This includes software as a service (SaaS) enterprise solutions, which provide extensive functionality from which organisations can select and pay for what they need, including services such as customisation and integration with other system components.

- **Generic application**: an application that is used as a tool for multiple purposes such as office productivity or communication.

- **Application platform**: usually considered part of the infrastructure domain; a collection of technology components (hardware and software) that provide the services used to support one or more business and generic applications.

Fallowdale Hospital: types of application related to patient communication solution

Application components include:

- **Business applications:** patient records system, outpatients scheduling system.

- **Generic applications:** spreadsheet (used for ad hoc prioritisation), email (used for interdepartmental communication).

- **Application platforms:** database management system (used for data storage), enterprise service bus (used to integrate multiple applications).

2.7 INFRASTRUCTURE ARCHITECTURE

Infrastructure includes primarily technological components and services that support the business and its activities. This can include tangible assets such as computer hardware, buildings and transport networks, and intangible assets such as software and supplier contracts.

Infrastructure architecture: The architecture of the technology infrastructure of an organisation or enterprise that supports the delivery of business activities.

Infrastructure includes hardware and software such as IT devices, operating systems, interfaces to non-IT devices such as vehicles, IoT and physical security devices, middleware, networks, communications, processing capabilities and standards. It is worth noting that **TOGAF** refers to this area as the **technology architecture**, **DoDAF** places infrastructure in the **systems view**, and the **Zachman** framework has it in two rows called **technology physics** and **component assemblies**.

The scope of infrastructure architecture is the same as for enterprise architecture, usually covering an entire organisation. For very large organisations, infrastructure architecture may be managed separately for autonomous subdivisions such as geographical regions, but there is usually some overall model that unites them.

2.7.1 Objectives

Infrastructure architecture activities aim to maximise the effectiveness and efficiency of the provision and use of infrastructure within an organisation or enterprise. This is largely driven by two factors: the changing nature of the business and developments in technology.

Changes to the business may include restructuring, such as mergers, as well as changes in business strategy in response to market conditions. Technological developments could include the availability of new hardware or software. Regular hardware and software life cycles can also provide an opportunity for analysis and evaluation of new technologies. These changes mean that infrastructure architecture must work continuously to maintain the closest possible alignment of the infrastructure with the operational needs of the business.

Infrastructure is a business overhead and so it is critical to minimise the cost of its acquisition and maintenance. Equally, the operational value of infrastructure to the business depends on maximising the performance efficiency of each infrastructure component.

An important benefit of the enterprise-wide view of architecture is that seeing the big picture enables the provision of infrastructure services that match the non-functional requirements (NFRs) from multiple sources rather than just the solution being developed.

Conversely, infrastructure architects are constantly seeking to eliminate duplication and unnecessary provision of infrastructure components, again taking the enterprise-wide view, so no components are removed that have active dependencies.

Like all the other domain architectures, infrastructure architecture must always ensure compliance with enterprise architecture principle and policy directives.

2.7.2 Artefacts

Typical artefacts include:

- **Infrastructure technology catalogue:** a portfolio that classifies components by the service or capability each provides, as well as containing details of supplier, contracts, technical specifications and current deployments.

- **Technical reference model (TRM):** consists of a taxonomy of the terminology, components and conceptual organisation of the infrastructure, usually accompanied by a graphical representation to aid understanding and communication. The information from a TRM may be made available through a configuration management database (CMDB).

- **Standards reference catalogue:** lists and describes all technical standards in use.

- **Hardware configuration view:** shows current (physical) and preferred (logical) configurations of infrastructure components to support design and rationalisation activities and to resolve conflicts.

- **Application technology matrix:** cross-references between applications and the infrastructure services they use.

- **Platform model:** shows the technology platforms that support the operations of business applications and the management of data.

2.7.3 Activities

Infrastructure architecture must support business activities and seeks to do so with maximum effectiveness and efficiency. This requires constant **rationalisation of existing components** and the design of an **ideal future state** for the infrastructure architecture of the organisation.

Solutions often require new infrastructure components and platforms and infrastructure architects therefore need to be aware of solution proposals and business cases. If a major change is being proposed, even if the infrastructure requirements are not known, an infrastructure architect might delay certain decisions around rationalisation.

When a solution design has been produced, the solution architect will make one or possibly a number of requests to the infrastructure architecture function to produce a **solution technology definition**. This is an assessment of the infrastructure that is required to support the proposed solution architecture. It can be produced as a high-level indicative statement to allow the solution architect to assess the **likely cost and impact** of the solution being proposed and perhaps decide to make changes before requesting a detailed report.

Infrastructure security auditing is a critical control that is part of an overall approach to managing enterprise risk. The frequency of auditing is dictated at a higher authority level by enterprise architecture or an enterprise risk management (ERM) function. Infrastructure architecture works closely with security architects and specialists to perform this function. Other security controls must be part of ERM such as monitoring of events, security alerts and testing, such as penetration testing and ethical hacking.

2.7.4 Relationship to solution architecture

Many infrastructure architecture components are invisible to business stakeholders even though they may help to enable the achievement of non-functional business requirements. Therefore infrastructure models and artefacts can be relevant as the basis for stakeholder views that are concerned with performance, reliability and other non-functional areas.

The infrastructure architecture function can advise on what, if any, changes are required to enable a solution to operate in terms of reconfiguring existing components, investment in capacity, scale, new hardware and software, and so on. This assessment provides essential details such as the capital and recurring costs of a solution and also an idea of the timescales involved.

A formal statement of the infrastructure required for a solution, including any changes to the architecture, is known as a **solution technology definition**.

Activity 2.4

A restaurant has identified the need for a way that customers can order food and drink from their table and pay the bill before leaving. The management is also considering providing a way for food and drink to be delivered to the table quickly and without any human waiting staff.

Identify some infrastructure services that will be required in these solution areas.

2.8 SOFTWARE ARCHITECTURE

Software is ubiquitous in the modern world and will therefore be part of many solution designs. Software components may be selected as part of a solution design based mainly on the functionality they provide, but non-functional aspects are also required to support business requirements. Software architecture therefore has the responsibility of making any software components comply with all requirements as well as conforming to standards and other enterprise and security directives.

Software architecture: the development and documentation of the structure and behaviour of a software system or component including its internal and external interfaces.

Note that a system operates at many levels, including:

- fine-grained components or units;
- integrated sets of units or subsystems;
- whole systems or applications.

Therefore the scope of software architecture can apply at the macro or micro level.

2.8.1 Objectives

The key objectives of software architecture are:

- **Design:** ensure software is designed and maintained in line with strategic directives.
- **Efficiency and effectiveness:** maximise the benefits of the functionality delivered by software.
- **Modularity:** provide flexible, reusable software components that build into subsystems and provide clearly defined services that are used as part of multiple applications and that can be enabled using techniques such as APIs web services and microservices.
- **Service delivery:** comply with SLAs and meet NFRs.
- **Maintainability:** minimise any constraints on change through well-designed software that is open to extension when business needs change.

2.8.2 Activities

Software architecture occupies the space between the applications and infrastructure architecture domains. Some software components provide the functionality that enables applications to operate and support business activities. Other software components deliver functionality that provides infrastructure services. In both categories are to be found some off-the-shelf systems and components and some that have been built to meet the specific needs of the business.

Software architects are often brought in to design software structures and interfaces to be built by developers as part of a solution design. However, requirements for designing new software and redesigning existing software may originate from either applications or infrastructure architecture when new or improved services are required in either of these domains.

Software architecture may also be consulted when a problem has been identified and the root cause is being sought. This is where an in-depth understanding of software architecture is used as the basis for analysing the problem and determining if the cause lies within the software.

Another situation in which the expertise of software architecture is required is impact analysis when a change is proposed, and it is necessary to assess the impact of the change on existing software.

Fallowdale Hospital: software architecture

The current letter-based communication system has a custom component that picks up messages to patients in a batch and delivers them as instructions to the letter printing, folding and enveloping equipment.

This component will still be required but interfaces will also be required to other communication channels. So far, text and email have been identified as required channels.

The solution architect has approached applications architecture to ask if such functionality is available within the hospital, perhaps as part of another solution.

Applications architecture has requested that solution architecture advise whether the existing component has the flexibility to work with other communication channels.

2.8.3 Artefacts

The artefacts of software architecture are, like other architecture descriptions, based around structure and behaviour.

The behaviour of software, especially modular software, is usually captured through descriptions of its interfaces. Where these interfaces are consumed by other software or systems, they are described in terms of **APIs**. These are commonly collected into libraries of software components which provide related functionality. Where the software interface is used by a person through a user interface (UI) then the functionality is captured with a variety of models such as the **unified modelling language (UML)** use case model. This shows the interaction between users or actors and units of functionality or use cases (Formal/2017-12-05, 2017). A **use case model** includes interactions with other internal or external systems.

The structure of software components is documented through **component models** that show how components work together and what data or other messages pass between them.

Since software works closely with data, either managing and processing it or using it to communicate, this data becomes part of software architecture and is documented using **data models**.

Requirements documentation is also relevant here so that the structure and behaviour of software components and systems can be traced back to the business requirements that they were designed to satisfy.

2.8.4 Relationship to solution architecture

The typical view of the relationship between solution and software architecture is:

- Solution architecture produces a design containing requirements that can be fulfilled using software.
- The software requirements are given to software architecture to design.

- The solution development team builds or acquires the software components following the design.

- The software and solution architecture functions govern the process.

However, there are a number of other points of contact between the two disciplines.

Solution architecture has multiple objectives that include producing the best solution but also minimising disruption to and negative impact on the business. This includes using existing components where possible. Therefore, when a design component is identified that could be implemented using software, the solution architect must find out if the component already exists.

One way to find out if a component already exists within the enterprise is by consulting the applications architecture portfolio. This categorises applications by the functionality they provide and maps them to the components that fulfil the functionality. However, the focus of an application catalogue is the functionality in use and existing software components often provide additional functionality that is not currently in use but could fulfil the needs of a solution.

In situations where additional or more detailed information about software components is required, the software architecture function can provide advice and analysis to establish the facts of the situation. This process may be led by applications architecture or through a direct relationship between solution and software architecture functions.

2.9 SECURITY ARCHITECTURE

Security is critical to all types and sizes of enterprise. Cybersecurity has the highest profile because the risks are greatest where organisations depend so much on connectivity and automation to deliver their business services. Many organisations are so dependent on technology that a serious attack could stop them functioning altogether, despite the fall-back or business continuity plans they have in place.

Security architecture is used to embed security design within the enterprise. Security cannot exist separately from the organisation or simply be an added layer. To be effective, the impact must reach all parts of the business.

Security architecture is distinct from security operations, which handles the day-to-day aspects of security such as auditing, testing, maintaining infrastructure components and dealing with incidents. Security architecture is about change and ensuring that when new infrastructure components, business applications and solutions are introduced, security is not compromised through poor design but rather maintained or strengthened.

Security architecture may be the source of change itself as part of the constant battle to stay ahead of new security threats. This means the design of security in all parts of the enterprise must be reviewed and maintained regularly and is likely to require change.

2.9.1 Objectives

The objectives of security architecture are to ensure that all architectural designs and components have appropriate security controls and measures built in. Specific objectives may vary by type of enterprise but include as a minimum:

- to minimise security vulnerabilities through design;
- to enforce security standards throughout the enterprise;
- to ensure all designs are robust against attack;
- to minimise the impact of security incidents.

2.9.2 Directives

Security architecture provides directives that are compatible and consistent with those of enterprise architecture and apply to all other domain architectures: business, applications, data and infrastructure.

For example, a principle that applies everywhere in the enterprise is that every asset must be assessed and classified. Assets are defined as things of value to the enterprise and therefore are vulnerable to attack or compromise that would have a negative impact on the enterprise. Assets exist as elements of all domain architectures and include:

- job roles (business);
- organisational units (business);
- information and data definitions (data);
- business systems and websites (applications);
- server nodes (infrastructure).

Other directives may be more specific, for example the classification of data and information assets by confidentiality, integrity and availability, which is mainly applicable to data architecture.

2.9.3 Activities

Security architecture is based on an assessment of the risks to an enterprise and its component parts, so this is a key activity for security architecture. It is an ongoing activity because both the enterprise itself and the environment in which it operates are changing frequently.

Security risk analysis and assessment must be performed every time there is a change to the enterprise architecture or subdomain architecture. This should be done in advance of any change by vetting proposed designs and reporting on their security issues with recommendations if appropriate.

Vetting of designs can take place at any level of granularity including enterprise, solution, software and component.

Security architecture must be continuously improved through rationalisation and consolidation of its components. This should occur in parallel with the assessment of change to the security situation.

Security architecture may be required to provide advice and guidance to individuals or groups such as architecture teams, design authorities and delivery teams.

2.9.4 Artefacts

Artefacts of security architecture include:

- Security principles, policies and business rules.
- Threat catalogue: details of risks against which assets are protected by controls.
- Control catalogue: details of controls such as access control, encryption and personnel screening.
- Asset catalogue: register of assets from all domain architectures and areas of the business.
- Risk analysis documentation.
- Data classification records

2.9.5 Relationship to solution architecture

Security architecture provides directives for all architecture subdomains that ensure all components are assessed for threats and suitable controls put in place to reduce or mitigate the risks. Therefore when solution architecture uses these components as part of a solution design, they should be secure.

However, within a single enterprise there are likely to be multiple controls of a similar category in use to protect different assets. When a solution design is produced, it must be assessed by security architecture to check that the controls are compatible with each other and that the design works as a whole.

Each solution has business requirements and some of these may relate to security. These requirements may affect some of the security aspects of the solution design. For example, a solution that combines information from multiple sources may have stricter confidentiality requirements than any of the individual data sources.

Security architecture may be consulted during the solution architecture life cycle to assess outline solutions and provisional designs and to compare alternative solution components. Security architecture is often invited, for example, to provide clarification of security aspects by a design authority.

REVIEW QUESTIONS

1. Which of the following is a subdomain of enterprise architecture?

    a. Strategic architecture.
    b. Applications architecture.
    c. Service architecture.
    d. Process architecture.

2. Which of the following categories of solution component are **not** from the business domain?

    i. People.

    ii. Organisation.

    iii. Process.

    iv. Information.

    v. Technology.

    a. i, ii and iii only.
    b. iii, iv and v only.
    c. i and ii only.
    d. iv and v only.

3. Grouping nurses, doctors and therapists together as a data entity called 'healthcare professional' is an example of what concept?

    a. Generalisation.
    b. Specialisation.
    c. Aggregation.
    d. Composition.

4. Why might previous solution designs be useful as inputs for a new solution architecture process?

    a. All necessary input artefacts can be obtained from existing solution designs.
    b. Previous solution designs show the best approach to solving the new problem.
    c. The problem space of the new solution may overlap with the previous solution designs.
    d. New solutions must always build on previous ones, even when addressing problems in a different area.

3 A FRAMEWORK FOR SOLUTION ARCHITECTURE

LEARNING OUTCOMES

When you have completed this chapter, you should be able to demonstrate an understanding of the following:

- The phases in the solution architecture framework
- The rationale for each phase
- The origin of solution ideas and concepts and the initiation of solution architecture
- Discovering and investigating architectural inputs, stakeholders and business requirements
- Developing a solution outline
- Business case management and option analysis
- The development of a logical solution design
- How the solution design is validated with stakeholders
- Producing and managing the solution design roadmap for delivery
- The physical design, delivery and governance of the solution.

3.1 THE PHASES IN THE SOLUTION ARCHITECTURE FRAMEWORK

The framework for solution architecture described in this book has been created specifically to complement the many published enterprise architecture frameworks currently in use. The framework incorporates techniques in common use by architects that have been tailored for solution architecture and organised in a sequential life cycle. Each phase of the life cycle has been designed to build on the outputs of its predecessors in order to achieve the best possible solution design for the business.

The process of producing a well thought-out, comprehensive and realistic solution design, starting with only the germ of an idea, requires skilled and organised work on the part of the solution architecture team. The framework for solution architecture provides a structure for the activities required to achieve a successful outcome, and organises them into a sequence or life cycle of eight phases (see Figure 3.1):

- **Initiation:** the starting point of the process where the business authorises solution architecture work to begin.

- **Discovery:** investigating the situation, engaging with stakeholders and gathering inputs.

- **Solution outline definition:** describing one or more solutions in high-level terms to stimulate feedback from the business.

- **Analysis:** deciding on one or two solution options to take forward and specifying the scope of change to be undertaken.

- **Logical design:** developing a single model of the solution and its components or building blocks.

- **Validation:** testing and assuring the design addresses the concerns of stakeholders whilst maximising positive impact and minimising disruption for the business.

- **Roadmap development:** producing a structured delivery plan with stakeholder priorities and timelines; this is a key document for the governance of the implementation of the solution.

- **Completion:** managed by programme and project management working with solution development teams and service management who will implement and deploy the solution, during which time solution architecture retains a governance role.

Figure 3.1 Framework for solution architecture

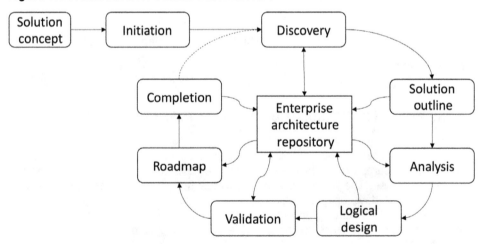

The process starts because the idea for a solution has been identified; this is known as the **solution concept**. Many artefacts will be used and produced during the process. Some of these artefacts will already exist and represent the current state; these are known as **baseline artefacts**. They exist within the enterprise architecture repository.

The solution architecture process will modify some of the baseline artefacts and produce new ones, such as **solution designs**. These modified and newly created artefacts are managed by solution architecture through the life cycle and the finalised versions are then lodged in the **enterprise architecture repository**.

3.1.1 Adapting the framework

Like many frameworks, the life cycle or **process model** of this one has several phases, each of which contains a number of activities. These phases and activities have been designed to work in a variety of situations, involving solutions of different types and sizes being implemented in a wide range of organisations.

When applying the framework to a specific solution or problem area, it is important to choose activities that make sense and to tailor them as necessary. However, even a solution that appears small in scope and simple in structure can benefit from following the steps in the framework consistently. This is true even if the work involved in completing some of the activities is small.

The same logic applies to the artefacts produced, known as the **content model**. The framework defines a number of artefacts or products that may be produced or modified during each phase and its activities. Some of these artefacts may need to be tailored for specific solutions, or to suit a particular way of working within an organisation or industry.

3.2 THE RATIONALE FOR EACH PHASE

The starting point for the life cycle is an idea, rather than a fully worked-out plan. Each phase develops the original idea, building knowledge and understanding to progress towards the best possible solution design that addresses as much of the problem as possible, whilst minimising cost and disruption.

Once the go-ahead is given to commence solution architecture activities, it is not inevitable that the life cycle will be completed. During or at the end of any of the phases, it is perfectly reasonable to abandon the process if no acceptable solution can be found.

Rather than stopping completely, it is much more common for the scope to be changed; for example, it can be reduced where only part of the problem can be addressed, or expanded to include parts of the business that were not originally included.

3.2.1 Initiation phase

This phase begins when the idea that a solution is required has sufficient support among senior stakeholders for action to be taken. The first task is to define and document the **problem area** in brief but sufficient detail to be understood by the stakeholders, who will decide on what happens next: whether to progress to the discovery phase or not.

The initiation phase should be extremely low-cost and not involve the extensive use of resources. However, the decision-making process requires senior business stakeholders' time and attention, so should not be entered into lightly.

Problems, risks and opportunities

The word **problem** is used in solution architecture to describe the situation that triggers the search for a **solution**. There are three broad categories of problem:

- Immediate problems or **issues** facing the business that must be dealt with.
- Anticipated problems that require **risk reduction** and mitigation measures.
- **Opportunities** that should be taken advantage of for the business to prosper.

These categories may be combined, for example when a business turns an issue into an opportunity.

A **problem space** delineates the scope of the problem and, in solution architecture, is the area of the business that must be investigated first to establish the root cause of the situation, and therefore determine what needs to change to provide a solution.

Once the problem area has been defined, it may be possible to identify a **business sponsor** – a senior stakeholder who will be the ultimate authority for the solution. The business sponsor may also be known as the product or service owner. The business sponsor is **accountable for the success** of the solution and makes all critical decisions, including whether to progress to the next phase of the life cycle. Other stakeholders may be informed and consulted as part of the decision-making process.

The outcome of this phase is a decision whether to progress to the discovery phase and, if so, how much time and resource is to be spent.

3.2.2 Discovery phase

Discovery begins with a well-formed description of the problem area, but this does not mean that the causes of the problem are well understood, or that there is there a clear idea about what to do to try to solve the problem. Even if suggestions have been made about the causes or potential solutions, a proper investigation must be carried out to establish the facts and ensure the correct action is taken to address the situation.

This phase focuses on three aspects:

- **Business requirements:** to establish what the business needs to happen in the problem area and how competing requirements can be balanced.
- **Existing architecture:** to try to understand the existing situation and start thinking about possible changes.
- **Stakeholders:** to involve the right people in the process and work towards a solution that is truly owned by the business.

At the end of the discovery phase, all the groundwork has been done and everything that is required to progress to a successful solution design has been gathered together. Through the process of identifying the causes and other factors in the problem area, stakeholders will start to have ideas and make suggestions as to what type of solution would be appropriate.

Based on this information, the business sponsor can make an informed decision as to whether solution architecture activities will continue for this problem area.

3.2.3 Solution outline definition phase

This phase focuses on producing one or more high-level solution designs. The product is called a solution outline, but is also known as the **conceptual design** because it is where creative ideas are worked through and modelled to see which ones are worth taking forward as candidates for a full logical design.

Concepts are key to keeping solution ideas at a high level and not limiting creativity and lateral thinking. Concrete solution components need to be made more abstract through techniques such as **idealisation**, **generalisation** and **composition**. Visual depictions and models are preferred, as they facilitate communication with stakeholders and will stimulate thoughts and ideas about the best solution.

As with any phase, the business sponsor may decide not to take any of the outlines or visions forward, either because they do not provide a satisfactory solution to the problem or because another course of action is preferred. Although a significant amount of work has been carried out to get to this stage, it is always better to stop or pause any further activity if there is doubt about the efficacy of a solution.

3.2.4 Analysis phase

Whereas the solution outline definition phase emphasises creativity and participants try to keep an open mind, the analysis phase seeks to narrow down the options and allow stakeholders to decide on a single solution design that will be developed and implemented. Note that it is possible to take two or more proposed solutions through to the next stage of logical design. However, the amount of work involved weighs in favour of choosing a single option. Either way, any decision is made by the business sponsor in consultation with other senior stakeholders, based on the best available evidence.

The solution options are presented as part of the business case so that they can be compared on equal terms and the correct one chosen, based on the benefit to the organisation or enterprise.

The selected option is then further refined by having its scope analysed and defined. Next, each building block that falls within the scope is identified and described in sufficient architectural detail (structure and behaviour) so that its role in the solution is clear and unambiguous. Once the solution components are in place, the interfaces between them are catalogued and described so that the content of each interaction (message) and its role in the solution are known and understood.

The output of this phase is therefore a single solution option (sometimes two or three) together with a catalogue of in-scope solution building blocks (SBBs) or components and the interactions between them defined as interfaces.

3.2.5 Logical design phase

Logical design brings together the outputs of the analysis phase into a single model. The structure and format of the design can vary but is usually based on a component diagram, with SBBs shown as components and the interaction between them shown as interfaces. The design must clearly represent all aspects of the solution, including organisational changes, information and data use, business activities and processes, and the role of technology.

The design produced is described as logical because it deals with logical components and their interactions, rather than the physical implementation of the design. Rather like an architectural design for a building, it is still a specification that will require some engineering work in order for it to be physically implemented.

3.2.6 Validation phase

Stakeholders have been fully involved in the solution architecture process up to this point, providing input and feedback and being involved in making decisions. This should mean that the design produced satisfies the need of the business, the wider organisation or enterprise, and all of its stakeholders. This includes:

- meeting business requirements;
- addressing the concerns of stakeholders;
- solving the problem;
- minimising disruption;
- minimising negative impact.

This phase aims to check all of these aspects of the solution in a formal and transparent way using three key techniques:

- **Views and viewpoints:** a viewpoint is a template that is applied as a filter to an architecture model with the result being a view. If the same viewpoint is applied to the before and after architecture description, then the effect of the change can be seen from a focused perspective by comparing the two views. Each viewpoint and the resulting views are designed to address the concerns of a stakeholder group.
- **Impact analysis:** a technique that examines each building block or component in a solution design and itemises the impact of change. Impacts can then be assessed as to their scope (reach), scale (size) and direction (positive or negative).
- **Gap analysis:** this compares two architecture descriptions and identifies the differences between them by listing all the components that are either added, removed or modified. Each changed component can be further described by the size and nature of the change, which can include, for example, estimates of resources and costs.

All three of these techniques are essential to confirm the suitability of the solution design but can also identify issues that, if considered serious enough, require the design to be reworked. If changes to the design are made, then all or part of the validation phase will need to be repeated.

3.2.7 Roadmap development phase

This phase begins with a fully worked-through design that has been tested and assessed against multiple criteria using a variety of techniques. This thorough approach should leave little room for surprise. The changes that have been identified, chiefly through gap analysis, need to be planned out and organised into a sequence that works from a **project management** perspective and is acceptable to the stakeholders in terms of when the benefits will be delivered. The other aspect of delivery that needs careful planning and assessment is the **management of risk**. The results of the impact analysis that was performed in the previous phase will be a useful resource, as it identifies and provides data about the risks due to the impact of change.

The output of this phase is an authorised delivery roadmap that has the approval of all stakeholders. The roadmap is authorised by the business sponsor. The roadmap is a **high-level plan**, and project and programme managers will need to produce more detailed plans during this phase to check that the solution can be delivered in the **sequence** and **time frame** of the roadmap. All project and programme plans that will be used to deliver the solution in the next phase, completion, must be consistent with the roadmap. The roadmap is the chief tool used for **communication with stakeholders** through to the delivery of the solution.

3.2.8 Completion phase

This phase aims to turn the logical solution design into physical reality, following the sequence of the delivery roadmap. Multiple disciplines will be required to carry out the work of **implementing the design**, including:

- Business analysis: changes to processes.
- Software engineering: bespoke systems.
- Technical specialists: hardware components.
- Information and data design: data storage and messaging.
- Human resources (HR): changes to job roles and skill sets.
- Business change: coordination of organisational transformation.

Solution architecture may be consulted at various points throughout the completion phase and has an important role in the governance of the delivery of the solution. Any major design decisions must be validated by solution architecture and an assessment made as to any wider impacts on the architectural landscape. Consultation may also be required to clarify solution requirements.

Solution architecture has a specific engagement with infrastructure or technical architecture for the process of solution technology definition. During this engagement, all the physical aspects of technology and technological services that support the solution are designed and agreed. There is a particular focus on fulfilling **NFRs** and defining **SLAs**.

The architectural output of this phase is a set of architectural artefacts representing the deployed solution in its entirety at the conceptual, logical and physical levels. These artefacts must be validated and signed off by solution architecture. They will then be lodged with enterprise architecture to become current state models in the enterprise architecture repository or continuum.

3.3 THE ORIGIN OF IDEAS AND THE INITIATION PHASE OF THE LIFE CYCLE

The impetus for change may be internal or external and is known as a **driver** (see Figure 3.2).

External drivers come from the environment in which the organisation or enterprise operates. These may be discovered by proactive investigation, for example using analytical techniques such as PESTLE (Paul and Cadle, 2020) or Porter's five forces (Porter, 1980). If so, then the organisation or enterprise usually has more control over the resulting business change, allowing more time for planning and contextualising the solution. If not, then the response is bound to be more reactive.

Internal drivers come from sources such as:

- business strategy;
- IT strategy;
- business analysis techniques, such as the life cycle for business change;
- enterprise architecture, such as the TOGAF Architecture Development Method (ADM).

Internal and external drivers may be reactive, proactive, or a combination of the two. For example, an unanticipated event may alert business owners to a situation of concern that may be further investigated proactively before any decision to act is taken.

Any documentation from formal analysis techniques is useful for solution architecture and may result in less work during the discovery phase.

If the decision is made by the business to proceed to solution architecture, then the next step is the initiation phase.

3.3.1 Solution vision statement

The first artefact that must be produced is a structured definition of why a solution is being considered. It needs to be documented adequately with three main sections:

- Description of the problem, risk or opportunity.
- Vision of the solution in outline with ideas or concepts.
- Ownership details and other stakeholder information.

Figure 3.2 External and internal drivers for change

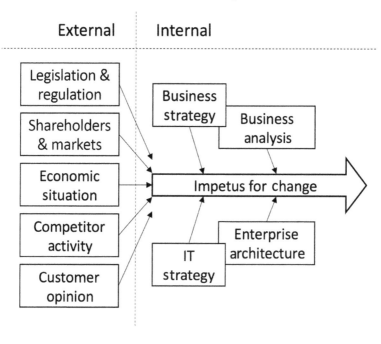

Various techniques may be used to make the description as complete as possible without spending too long on analysis. One such technique is based around the six basic questions: why, how, what, who, where and when.

Fallowdale Hospital patient communications: problem description

What is the problem? Patient communication by post is expensive, ineffective and potentially insecure.

Why is this a concern? The hospital needs to reduce costs, improve performance and avoid breaching data privacy laws.

Where does the problem occur? In the post room where letters are printed and dispatched, in hospital Outpatients where patients attend clinic, in community locations where outreach clinics are run, in the call centre where patients call when there is a problem, in clinics where staff deal with non-attendance and in internal investigations when a patient suffers medical harm.

Who is impacted? Patients, Outpatients admin, clinicians, directors (medical, nursing, operations, finance) and post room staff.

When did this start? When the first centralised printer and letter folder was introduced 27 years ago (it was replaced 12 years ago by a more modern version), although this was an improvement on the previous process (using standard printers). Since then, costs have steadily increased (licence fee, maintenance, staff, materials and postage). The local population has grown. There are many more people in rented accommodation, including an increase in student numbers, meaning people change address more frequently.

How important is this? Patient communication has been identified as central to the delivery of care, and addresses a number of other business objectives such as equal access to care. Outpatient clinics deal with more patients than any other part of the hospital and improvements in efficiency here will have a large effect on the capacity for treating patients.

Activity 3.1

From the Fallowdale Hospital patient communications problem description, identify some values that can be measured to quantify the problem now and then measure any improvements.

It is very tempting to start thinking about some specific ways of solving the problem, but this should be avoided at this early stage. It is better to know all the facts before jumping to a conclusion. However, it is important to have a vision of what the solution will do for the business and other stakeholders – in other words, the effect of solving the problem.

Whichever person or group identified the need for a solution undoubtedly has some ideas about their vision of the future. These need to be separated from any detailed suggestions on how to solve the problem. A technique to expand on these ideas is to ask those involved to describe the situation as if the problem had been solved already. This can be done using storyboards and scenarios to represent the patient journey.

Fallowdale Hospital patient communications: solution vision

- Patients always attend appointments.
- Patients know exactly what to expect when they attend.
- Appointments run smoothly.
- All necessary information is available when needed.
- If a patient cannot attend, they contact the hospital and alternative arrangements are made.
- Patients are fully informed after the appointment.
- Relevant healthcare professionals are fully informed about the patient's care (GP, other consultants, inpatient ward).
- Patient contact details are always current and correct.
- Data breaches concerning patient data never occur.

Successful business change requires leadership, so it is important to find someone from the business who can take ownership of the solution and make sure it happens. This person will have the most interest in seeing the solution in place, so it is usual to choose the person who will benefit most when the solution has been deployed.

There are always multiple stakeholders for any solution. Some key stakeholders will be identified during the initiation phase, although a complete stakeholder catalogue will not be available until the discovery phase.

Fallowdale Hospital patient communications: stakeholders (initial list)

- Patients.
- Outpatient administration.
- Clinicians (consultants and those running clinics).
- Medical director.
- Director of nursing.
- Operations director (identified as potential business sponsor).
- Finance director.
- Post room staff.
- GPs and community staff such as district nurses.
- Diagnostic staff (tests are coordinated with appointments).

3.3.2 Architecture initiation document

This is the document that, once authorised by the business sponsor, gives authority to the solution architect or architecture team to proceed to the next phase in the life cycle: discovery. The architecture initiation document is a contract between the business, represented by the business sponsor, and the solution architecture function who will deliver the architecture work. This document performs a similar function to the **architecture contract** in the TOGAF ADM.

The main sections of the document are:

- **Stakeholders:** also included in the solution vision statement, but here specifically identifying the business sponsor and listing the key stakeholder contacts that solution architecture should begin with.
- **Time frame:** during which the tasks of the discovery phase are to be completed and any relevant events occurring during this period, such as meetings where stakeholders may be consulted or updated on progress.
- **Resources:** that can be dedicated to the solution architecture activities, including people, equipment and money.

- **Deliverables:** to be produced during the discovery phase. The type and level of detail of these outputs must be aligned with the resources allocated and the scope of work that is being authorised.

- **Scope:** of business activities that are to be investigated.

- **Constraints:** that must be observed during the discovery phase in addition to time and resources.

- **Terms of reference:** for the solution architecture work, including who will lead the work, arrangements for reporting back to the business, escalating issues and notifying risks.

- **Authorisation:** of the document and the instructions it contains by the business sponsor or another senior stakeholder.

3.4 DISCOVERY PHASE

This is a critical phase of the solution architecture life cycle because it creates a solid foundation for designing a successful solution. This foundation comprises three parts, which are matched by the three activities of the discovery phase (see Figure 3.3):

- **Facts** about the current situation that are relevant to the problem area, collected in the **architecture inputs** activity.

- **People** who can help to make the solution fit the business, assessed in the **stakeholder engagement** activity.

- **Criteria** that the solution must fulfil in order to be successful, catalogued in the **business requirements** activity.

The essential **facts** can be obtained in a usable form from existing architectural artefacts and descriptions that should be available from the enterprise architecture repository. Some information may not be available and will require investigation and modelling to make it useful for solution architecture. Some existing architecture artefacts may require modification or reformulating in some way before they can be used.

The **people** who will contribute to the design through their specialist knowledge and responsibilities for decision making within the organisation or enterprise are the solution stakeholders. Some will be known already, especially as there is a list provided as part of the solution vision statement that is an input to this phase. Other stakeholders will be discovered as part of the stakeholder engagement activity.

The **criteria** for the success and quality of a solution are expressed in terms of business requirements. Some requirements will already have been identified as part of the solution vision statement. Others may be inferred by analysis of the problem using existing or newly created architecture artefacts. Additional requirements will be identified by stakeholders and all need to be carefully documented and managed.

Much of the work in the solution architecture life cycle occurs in parallel and is often revisited and refined as the life cycle progresses and more is known. This includes the three activities of the discovery phase. It should be noted, however, that business requirements come from architecture inputs and stakeholder engagement. Similarly, stakeholders can be derived or inferred from the existing situation which is embodied in the architecture inputs. Therefore the sequencing of activities in the discovery phase can be understood as representing the natural order of progression, from the discovery of facts and the understanding of the problem to what needs to be done so that it may be solved.

Figure 3.3 Activities of the discovery phase

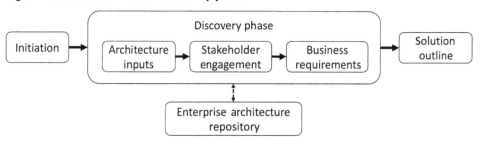

The discovery phase sits between the initiation phase, when the problem area has been described in some detail, and the solution outline definition phase, where the basic elements or concepts of the solution are defined and modelled at a high level.

3.4.1 Architecture inputs

A number of factors determine whether useful architecture artefacts are available at the start of the discovery phase:

- **Architecture maturity:** whether architecture practice is established and working well in the organisation or enterprise and if there is an efficient repository of artefacts that can be easily accessed and in which it is easy to find relevant artefacts.

- **Overlap with current problem area:** if the new problem to be solved is in an area or operational domain that has never been addressed before, there will be fewer relevant artefacts.

- **Clarity of understanding of the current problem:** if the focus of the problem is too broad or ill-defined, it is harder to distinguish relevant artefacts and too many irrelevant ones may be included as inputs.

The main product of this activity is a catalogue of relevant architecture artefacts for the operational domain of the solution vision. At this stage it is not necessary to examine the artefacts in detail, but it is important to identify them and know they are available when needed.

If the artefacts are organised in a repository then more details will be available, such as:

- date of production and modification;
- version number;
- owner;
- business area;
- related artefacts that would be impacted by change and any dependencies.

It is also useful to record the reason for selecting this artefact by linking it to an element or concept in the problem domain.

For planning purposes, it will be necessary to know if an artefact requires any significant rework or completely new artefacts created, for example because connected areas of the architecture have changed since it was produced. However, it is critical to limit the amount of work done in the discovery phase because of budget and resource constraints. It is never a good idea to undertake work just because it might be useful in the future, rather it is better to do the work just in time for when it is needed. This is especially true with architecture work where there are so many interdependencies, meaning it is highly likely that something will change and that work will have to be redone.

3.4.2 Stakeholder engagement

Solution architecture is concerned with stakeholders for the current solution.

 Solution stakeholder: an individual, group or organisation having an interest or concern in a solution.

This definition has been adapted from ISO/IEC 42010:2011 and TOGAF.

Examples of solution stakeholders are:

- Customer (internal or external).
- Business owner.
- User or business actor, involved in the operation of the solution.
- Consumer of a product.
- Supplier of a component of the solution.
- Business partner, external to the organisation but involved in the solution.
- Designer.
- Maintainer.
- Subject matter expert (SME) for the problem or solution domain.
- Architect.

Stakeholders can be identified from a number of sources:

- **The solution vision statement:** contains a provisional list of stakeholders identified during the initiation phase.
- **Stakeholder registers for similar solutions:** may contain some stakeholders for this solution.
- **Architecture inputs for this solution:** these are artefacts that describe the problem domain, from which can be derived stakeholders with responsibility or involvement in that domain.
- **Investigation of the problem domain:** leads to the discovery of more stakeholders when any individual, group or organisation is found to be involved with or impacted by a solution in that domain.

The primary artefact produced by this activity is a **stakeholder register**, identifying potential and confirmed stakeholders for this solution together with their roles, responsibilities and concerns.

Activity 3.2

Think of one possible artefact from each of the business, data and applications architecture domains that are relevant to the Fallowdale Hospital patient communications solution.

From the artefacts you have selected, can you now identify any related stakeholders for the solution?

A secondary artefact that can be produced at this stage is a **stakeholder communication plan** showing how and when each stakeholder group will be involved in the solution architecture process.

3.4.3 Business requirements

Business requirements represent the **criteria by which a solution will be judged**. They act as a measure of quality; if all requirements are met, then the solution is of high quality. Of course, there are also project constraints, such as those of time and resource, by which the success of the solution delivery may be measured. A solution that perfectly delivers its business requirements will still be judged a failure if it vastly exceeds budget or misses a key deadline. The responsibility for delivery within these constraints falls to P3M, but they should be documented here so that they can be included in the roadmap development phase.

Requirements are also the **basis for designing a solution**. New requirements are often discovered during the design process and may appear at any point during the solution architecture life cycle. However, the earlier they can be discovered and evaluated the better. Solution designs consist of highly interconnected components and a new requirement being introduced after some or all of the solution has been designed will almost certainly mean reworking the design.

It is important therefore to deal with requirements in an efficient and effective way so that the design process can progress towards a high-quality solution. Important activities include:

- **Capturing:** documenting business requirements as simply as possible, but in sufficient detail so they are clearly and unambiguously understood and are usable for the purpose of designing a solution.

- **Validating:** checking they are complete and correct and represent what has been asked for by the business.

- **Verifying:** testing that they are genuine requirements for this solution and that they do not conflict or overlap with other requirements.

The process of handling business requirements can be strengthened using techniques such as those in a requirements engineering framework (Kotonya and Somerville, 1998; Sutcliffe, 1996).

The key product of this activity is a **requirements catalogue** that itemises requirements and gives basic details such as:

- name;
- ID;
- description;
- justification;
- category;
- status;
- change control details.

A requirements catalogue is useful as a central point of reference and can be used to collect requirements before any further processing. This artefact also serves as a way of tracing business requirements that may perhaps become more detailed system or data requirements, or may be linked to operational measures such as SLAs.

Other artefacts may be used in addition to a requirements catalogue, either to provide more detail, if required, or as part of the process of verification where diagrammatic modelling techniques are often used.

Artefacts such as **use cases**, **user stories**, **data models** and **business scenarios** are often used for capturing the details of requirements. These are also used by other techniques and methodologies, including Agile, and are consistent with enterprise architecture.

3.5 SOLUTION OUTLINE DEFINITION PHASE

The aim of this phase is to develop a description and illustration of one or more potential solutions that satisfy the business requirements and therefore solve all or part of the problem. This is challenging, because it requires a balance between showing stakeholders an approach that they think will work without unnecessarily constraining the final design. In other words, it is important to keep an open mind and not jump to conclusions about the type of solution. There is more investigation and design work to be done before the final solution architecture is produced. This work could be hampered if inappropriate decisions are made that constrain the solution design before proper analysis has been completed.

The solution outline or outlines produced at this stage therefore need to be at quite a high level. This is known as a **conceptual design**, because it is at the higher end of the **idealisation hierarchy** that runs from conceptual through logical to the lowest level, physical.

More than one outline definition may be needed if there is unresolved disagreement among stakeholders. Differences can often be resolved during this stage by raising the level of abstraction of the outline to a level where everyone can agree.

3.5.1 Visual representation of the solution

Most solutions have multiple stakeholders with different perspectives on the situation and concerns about how it should be dealt with. It is important to get input and feedback from all stakeholders and ultimately reach agreement on how to proceed.

This type of multi-party collaboration depends on the ability to communicate easily and to use multiple media. Visual representations are an excellent way to convey complex ideas and designs and can be easily communicated in multiple ways, such as:

- presentation to a large group of stakeholders;
- web page with links;
- document via email;
- briefing notes.

A visual representation that is suitable for communication to all stakeholders must be as simple to understand and as abstract as possible to limit the amount of specialist knowledge required to interpret it. There are multiple forms of visual representation that can be used here, such as:

- context diagram;
- rich picture;
- component diagram.

3.5.2 Supporting artefacts

The high-level visual representation of the solution needs to be supported by more detailed information in order for different stakeholders to understand how it addresses their area of the business. One way of doing this is to expand the virtual representation of the entire solution into multiple focused diagrams.

Other artefacts can be used to provide extra information or explain certain aspects of the solution. These include:

- tables;
- cross-reference grids;
- maps;
- storyboards;
- business scenarios.

These diagrams can be backed up with text-based artefacts to capture even more detailed information.

3.5.3 Traceability grid

An artefact that is very useful for the management of stakeholders, their concerns and business requirements is a traceability grid. This is a simple cross-reference grid showing the links between solution design elements and the original solution vision, including an explanation of any variations.

Many variations from the original concept of the solution occur because of the identification of requirements or stakeholder concerns during the discovery phase. Each newly discovered requirement could lead to a new or modified design element in the solution outline definition.

Capturing these links in a transparent way means that there is a much greater understanding among stakeholders of how the solution design was created and why the design decisions were made.

3.6 ANALYSIS PHASE

The analysis phase takes the high-level design artefacts produced at the solution outline definition phase and looks at them in detail and from all angles to prepare for the construction of the logical design (see Figure 3.4). This includes considering a range of options for the solution approach that the stakeholders can assess and select, then formally scoping the solution and looking in detail at potential components and all solution interfaces, external as well as internal.

3.6.1 Business case

This activity aims to put forward a brief description of a range of options, including sufficient detail for stakeholders to be able to select the best one to take forward. This is not always a straight choice between several distinct options; it may well be possible to choose one option but request that it incorporates some of the benefits of another. This may then require further investigation to assess the feasibility of such a course of action.

Each option that goes into the business case selection process has a:

- description;
- assessment of scale of change using gap analysis;

- cost benefit analysis;
- return on investment report;
- risk assessment.

Figure 3.4 Activities of the analysis phase

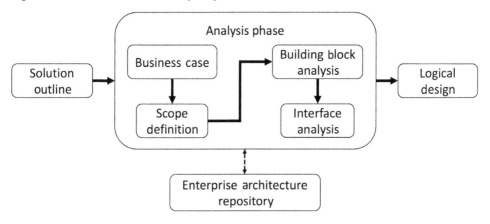

During the selection process a trade-off analysis and overall risk assessment are prepared and documented with input from stakeholders.

At this early stage in the analysis phase many of the details have not been worked out, so only a high-level outline of each option is presented to stakeholders. One of the key factors in the decision-making process is the scope and scale of the change. This gives an indication of the effort involved and the impact on different parts of the business. It should be possible to give a broad indication of the business units, processes and services that will be affected. The scale of change is harder to quantify without further analysis but can be estimated.

A preliminary risk assessment should be completed and the findings attached to each option. The benefits of each option should also be itemised. Any known costs can be included, but it is too early to perform either a complete cost benefit or return on investment analysis.

It may seem strange to ask stakeholders to make a choice based on such limited information. The alternative would be to carry out a full analysis on all options. This would be much more resource-intensive, costly and time-consuming. It would also mean that stakeholders would not have given a clear indication of their preferences regarding the solution. Therefore it is better to have the option selection at this stage and focus the effort on a solution which stakeholders have already considered and provided their input.

Although all stakeholders can provide input during business case option selection, the business sponsor has the final say and must authorise the option that is taken forward.

Activity 3.3

One of the senior stakeholders for the Fallowdale Hospital patient communications solution has suggested that communication with patients could be managed using a cloud-based CRM system.

From the information provided so far, prepare this approach for consideration as a business case option.

3.6.2 Scope definition

The business case gives a broad indication of scope for each option under consideration. Once a choice has been made and any modifications agreed, the solution scope can be formalised.

A solution scope definition should itemise each area of the business where changes will be made. This is likely to include:

- business units;
- job roles;
- business processes;
- products and services;
- IT systems and infrastructure.

The scope should indicate whether the changes will involve new or existing items. For existing items, the scope should indicate if the item is to be modified, removed or replaced. Note that replacement of an item is similar to a combination of adding a new item and removing the existing item, but with the additional constraint that the interfaces must be maintained or enhanced.

Each item or area in the scope definition should be linked to any architectural documentation available. This is to be found in the enterprise architecture repository. Any additional architecture documentation that has been produced specifically for this solution should be registered with the enterprise architecture function. If architecture work is unknown to the enterprise architecture function, there is a risk that overlapping or conflicting work may be initiated elsewhere in the organisation.

The scope definition is a critical document for many of the future activities throughout the solution architecture life cycle.

3.6.3 Building block analysis

SBBs are the components of a solution. Building block analysis aims to examine each component in the solution in turn to see if it exists already in the organisation and if not, whether other components can provide all or part of the necessary functionality. The list of building blocks can be constructed by combining the items in the scope definition document with the areas from the conceptual design.

For any building blocks that already exist, the solution architect should make an assessment of whether they are acceptable, ideal, replaceable with the same interfaces, duplicatable with a plan for future replacement, or need to be built (or sourced) from scratch.

The output of this activity is a solution building block model that lists SBBs and gives details of:

- name;
- category;
- ownership;
- current state;
- required modifications.

The catalogue may be tailored to include additional information that is useful to the organisation or enterprise.

3.6.4 Interface analysis

At this stage there should be an emerging picture of the SBBs and how they will work together. The first task here is to construct a list of all the touch points, hand-offs or interfaces between the SBBs.

Each interface then needs to be described in terms of:

- source;
- destination;
- trigger and other events;
- items exchanged, including any supporting information;
- sequence;
- pre- or post-conditions.

The output of this activity is a solution interface catalogue.

3.7 LOGICAL DESIGN PHASE

The logical design phase brings together the outputs of the analysis phase into a single model. The model structure and format can vary but is usually based on a component diagram, with SBBs shown as components and the interaction between them shown as interfaces (see Figure 3.5).

Figure 3.5 Logical design

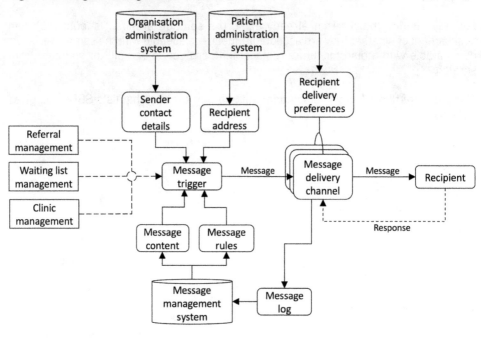

The scope should match that of the scope definition document and any design decisions that represent a change in benefits, or anything agreed in the business case, need to be itemised and accompanied by explanatory notes. All aspects of the solution need to be clearly represented here, including organisational changes, information and data use, business activities, use of technology, and so on. Ideally any KPIs for components or interfaces should be shown either on the model or in supporting documentation.

i

Design patterns

Design patterns are a concept that originated in the architecture of towns and buildings. In *A Pattern Language* (Alexander et al., 1977), architect Christopher Alexander describes a pattern as 'a problem that occurs over and over again in our environment, and [...] the core of the solution to that problem'. This definition was quoted in one of the most famous books on the subject within computing, *Design Patterns: Elements of Reusable Object-Oriented Software* (Gamma et al., 1994).

Alexander gives many patterns related to towns and buildings, such as independent regions, web of public transport, road crossing and main entrance. Some of the patterns are more abstract, such as a place to wait and accessible green.

These patterns embody experience and even wisdom that can be passed down the generations and can include ready-made solutions to common problems. For example, in a town there may be many road crossings either by pedestrians or by other roads. Ready-made solutions for pedestrians include zebra crossings, pedestrian traffic lights and foot bridges. For roads there are crossroads, traffic lights and roundabouts. A pattern may give additional guidance on when to use these various alternatives.

Patterns can help to recognise a problem and provide solutions, but can also spot gaps. For example, a place to wait is often forgotten about in designs for buildings and towns. Having a list of patterns is a useful reminder of the common components that should at least be considered for inclusion in a design.

Solution architecture is not as mature as the architecture of towns and buildings, but there are many recurring problems and solutions. For example, identity and access management is a common component of problems in various domains. Components that can be used to address this in a solution include identity cards, RFID, certification authorities, retina scanning and document classification.

Other solution architecture patterns could include multi-party collaboration, two-way communication, distribution of scarce resource.

Perhaps one day there will be a book about solution architecture patterns that is as instructive as the two mentioned above; however, over time new patterns emerge and existing ones are modified, and the practice of solution architecture allows the lessons of change to be captured and used to improve future designs.

Other types of design pattern are relevant to solution architecture, including:

- data design patterns;
- enterprise integration patterns;
- business model patterns.

If the design is for a modification to an existing solution, then the logical design can be presented alongside any existing design documents for the current state solution.

3.8 VALIDATION PHASE

A certain amount of validation will have been completed during the process of producing the logical design. It may well have been shared with selected stakeholders for their ad hoc opinions and more formal input during construction. Where there are design choices to be made, some of the most relevant of the enterprise architecture models may be consulted to assess the best path to take.

The validation phase formalises the process of ensuring that the interests and concerns of stakeholders are being accommodated and balanced and that the design achieves the maximum positive impact and minimises unplanned disruption (see Figure 3.6). Note that planned disruption may be desirable and beneficial to the business. In short, this phase aims to validate that the solution architecture is designed in the best interests of the business.

Figure 3.6 Activities of the validation phase

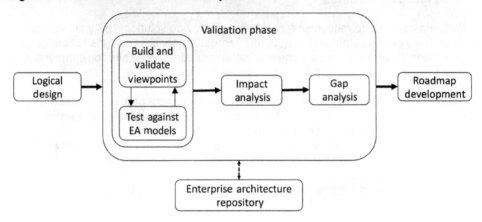

3.8.1 Build and validate viewpoints

This activity looks specifically at stakeholder concerns. Some will have been identified during stakeholder engagement during the discovery phase. Other concerns become apparent through the ongoing contact with stakeholders as the shape of the solution design emerges and evolves. Each concern must be documented and recorded in the stakeholder register.

It is prudent to hold a review of the stakeholder register before starting to build and validate viewpoints. The review can check if all stakeholders are included and whether they have concerns about the solution. This can be verified with stakeholders to ensure that all concerns are known about, up to date and correctly documented.

Multiple stakeholders may share the same concerns and so the list can be rationalised to group concerns into similar areas that can be addressed together. The next task is to define a number of viewpoints to show the current and future state of the area of architecture covering each group of stakeholder concerns. A single viewpoint will be used to produce multiple views of the areas of the organisation affected by the solution. There will be at least two views produced from each viewpoint, representing the current state and the position after the solution has been implemented. Additional views will be needed if there are transition states on the way to delivering the solution.

Viewpoints specify two aspects of the views that will be produced. The first is the list of architectural elements that will be selected for the view and about which information will be presented to stakeholders. The second is the visualisation that will be used to present the information. The form of presentation used needs to fit the information being presented and address the concerns of the stakeholders who will be the view's audience. This could be a table of data, graph, map or any other diagram (or combination of these artefacts) that successfully addresses stakeholder concerns.

Viewpoints are tested by producing views and checking that the information presented matches that in the underlying architecture. Viewpoints and the views produced need to be registered and lodged in the enterprise architecture repository for future reference.

When the views are presented to stakeholders, there may be further work to do. For example, some stakeholders may not be convinced that their concerns are being addressed. This may require reworking of the viewpoints to show some missing aspects of the solution. If the concern has genuinely not been addressed by the design, parts of the solution may need to be redesigned. If the solution's architecture changes, this could potentially affect all views produced by all viewpoints, so these need to be rerun in order that all stakeholders can check that their concerns are still being addressed satisfactorily.

When stakeholders are happy with the proposed solution design and the resulting future-state architecture, this fact needs to be recorded either through the minutes of an approval meeting or as a formal sign-off.

3.8.2 Test against enterprise architecture models

The solution architect must ensure that the proposed changes will produce their claimed benefits, the evidence for which is obtained from modelling the change using models that should be available in the enterprise architecture repository. Ideally, solution architecture makes full use of models to analyse, design and test solution blueprints to ensure they are robust in themselves and fully integrated with other solutions and parts of the business.

In the previous activity, building and validating viewpoints, the concerns of stakeholders were used as the criteria for assessing the suitability of the design of the new solution. Many, if not all, of those concerns will be shared by the solution architect and some of the viewpoints may contain additional details to address concerns that go beyond those already identified by other stakeholders.

Modelling at this stage is a way of establishing if the proposed solution is the very best it can be in terms of providing positive benefits with the least change and disruption and the minimum negative impacts. It is always to be hoped, of course, that the negative impacts can be totally eliminated.

There are other possible considerations that can be tested at this stage. For example, the solution in focus may be part of a larger programme of change; this would demand that its solution architecture is fully consistent with other solutions and target architectures.

3.8.3 Feedback between activities

Modelling and testing against enterprise architecture models may identify changes to the solution architecture. If this is the case, affected views will need to be rerun and presented to stakeholders, and approval for the changes in the context of stakeholder concerns may need to be sought and given. Depending on the level of change to the architecture, some viewpoints may need to be altered to include the changes.

There is therefore a potential feedback loop between the activities of validating viewpoints and testing against enterprise architecture models. Sometimes this detailed scrutiny by architect and stakeholders may reveal a flaw or opportunity for improvement in the design, which necessitates returning to the logical design phase before repeating all the activities in the validation phase. This illustrates the iterative nature of architectural design.

3.8.4 Impact analysis

Impact analysis aims to identify and categorise all impacts of all the changes that will occur when the solution is implemented.

> **Zone of impact**: the area of the enterprise that will be affected, positively or negatively, by the implementation of a solution. Impacts may be **internal** to the solution, being within its scope, or **external**, where the impact may be in any part of the enterprise or beyond its boundaries.

In order to capture all the changes, it is important to have a complete and up-to-date scope for the solution. This was originally agreed during the analysis phase and may have changed since that point. Any changes should have been agreed and recorded, but it is worth reviewing the scope at this stage to ensure completeness, as it is a critical input to the activity of impact analysis.

The scope definition should link items to related architectural elements. This is also worth reviewing as the architectural artefacts contain valuable information to assess areas of impact.

The starting point for the activity of impact analysis is a complete list of areas of the business that will be subject to change as a result of implementing the solution. This includes direct and indirect change through interfaces. The enterprise architecture models should be used to explore the possibility of onward impact beyond any actual change.

Impact areas that are commonly affected include:

- capabilities;
- staffing;
- business services;
- business processes;
- relationships with external organisations.

Each impact area that has been identified is then assessed to see whether the impact is overall negative, positive or zero. The category and type of impact should also be assessed and recorded.

Impact categories that should be considered include:

- loss of revenue;
- increased operating costs;
- additional capital expenditure;
- penalties, fines and other sanctions;
- customer experience;
- impact on staff.

Some impacts have the effect of increasing, decreasing or mitigating the risk of something happening. Other impacts are more definite, and if the effect is negative they are considered as issues. An issue is simply a risk with 100 per cent certainty.

Arguably, impact and gap analysis can be carried out in any order or in parallel as there does not seem to be much of a dependency between them. However the impact analysis is more likely to lead to a revision of the solution design and target architecture. This could have a knock-on effect on the architectural artefacts that are compared during the gap analysis activity. Therefore, if possible, it is preferable to complete the impact analysis first.

3.8.5 Gap analysis

Gap analysis in solution architecture uses the architectural entities, artefacts and interfaces within various models to identify components and interfaces that are different between the current and future states. These are then itemised and categorised as either to be added, deleted or modified. Any component to be modified can be further broken down into its subcomponents and gaps identified at this lower level of decomposition.

Simply having a list of what needs to change does not give any guidance as to how to bring about the change, but the advantage of the architectural approach is that the changes are broken down to component level and categorised, which is a good starting point. A typical next step would be to assess the amount of activity, resource, cost, and so on, that is required to achieve the change to each item in the list. Additionally, factors such as risk and interdependency may be recorded and used as the basis for organising the activity into a project plan or roadmap.

Gap analysis has two main deliverables:

- **Gap analysis report:** itemises all the differences between the current state and the target state for the solution, broken down into the components or SBBs, and interfaces involved in closing each gap. For each change from current to target state, the report has an estimate of the effort (time and resource) and cost that is required to achieve it. Additional factors such as risk and interdependency can be recorded, if known, and these can be used as the basis for developing the delivery roadmap.

- **Gap models:** show the transition from current to target state using the artefacts including architecture models and views that have been used in the solution architecture process and are therefore familiar to stakeholders. Each artefact should show all the architectural elements in its scope, including those that will remain unchanged as well as additions, deletions and modifications, usually highlighted or colour-coded for clarity. These models are the basis for producing the gap analysis report and provide supporting documentation for it to enable stakeholders to drill down into the proposed changes.

3.9 ROADMAP DEVELOPMENT PHASE

The product of the roadmap development phase is the eponymous roadmap or high-level communication document, which aims to show the components of a solution with the focus on the business benefits or **capabilities** that will be achieved and the estimated **time** by which they will have been achieved (see Figure 3.7). Typically, this will be shown as a **timeline**, often broken into tranches. The purpose is to show stakeholders what will be achieved and by what date.

Figure 3.7 Solution roadmap

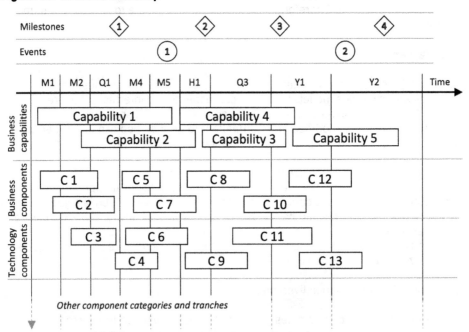

Much of the detail and complexity of project and programme planning is hidden behind the roadmap view, because it is less relevant to the stakeholders who are mainly focused on the needs of the business. However, this type of view does give the stakeholders sufficient information to judge whether the proposed timelines and benefits are adequate to meet their needs and will fit in with the timescales of enterprise planning.

The roadmap for a solution will require a great deal of input from portfolio, project and programme management (P3M) because they have responsibility for delivery. However, the solution architect who produced the logical design and other artefacts will also have a great deal of involvement. At this stage the solution architect is in a good position to represent the business with their clear understanding of the requirements, constraints, deliverables and dependencies. The P3M staff will be able to estimate and define the resources needed to deliver the components of the solution. The roadmap is therefore a collaborative deliverable.

Once a proposed roadmap has been constructed, it will be shared with the key stakeholders to be discussed, modified if necessary and then agreed as the way forward. It will remain the primary communication document throughout the delivery of the solution, used for updates and for discussing any proposed changes to timescales, priorities, and so on.

Fallowdale Hospital patient communications: roadmap elements

- Existing standard messages based on a template (changes made separately to each of 100+ templates).
- Configurable messages allowing rules to govern content.
- Integration with waiting list management processes (including add to list, remove, prioritise and suspend).
- Integration with referral management processes (including referral received, accept and reject).
- Integration with outpatient management processes (including attendance, non-attendance, cancel appointment and reschedule appointment).
- Responses by recipient (patient or service user) including accepting or rejecting appointment offer.
- Recipient to control methods of communication through personal preferences.
- Postal service delivery channel using printed letters (currently in use).
- Email delivery channel.
- Instant message delivery channel.
- Automated phone communication channel.
- Secure client website for patients, service users and external healthcare professionals.

Activity 3.4

Given the roadmap elements that have been identified for the Fallowdale Hospital patient communications solution:

    a.   List the items that must be delivered as a minimum set for the initial release.

    b.   Which other deliverables would follow in the next update?

    c.   Give a brief explanation for your choices.

3.10 COMPLETION PHASE, DELIVERY AND GOVERNANCE

After the solution design has been finalised and the delivery roadmap agreed, the process of implementation by solution developers begins under the control of programme and project management. This phase will include but not be limited to:

- software development and acquisition;
- business process change by business analysts;
- organisational changes, including the retraining of staff, which will most likely be implemented by HR working with business managers.

This phase turns the **logical design** produced by solution architecture into a **physical implementation** that can be deployed within the organisation or enterprise.

The modular design, which is the foundation of solution architecture, makes it highly compatible with many of the solution development styles and methodologies in current use, such as the incremental delivery of Agile and Scrum, and the object-orientation of languages, such as Java, Python and Ruby.

Once implemented, the solution will be validated by the business using acceptance testing techniques to check that all business requirements have been met. The solution architecture function is able to clarify business requirements for the solution and can verify that they have been met as part of the delivery governance process.

Acceptance testing is closely associated with software, but all parts of the solution need quality assurance. The solution architecture function has an advisory and governance role here as well.

The deliverables from this phase are the completed physical designs and configurations that are validated and lodged in the enterprise architecture repository or continuum as current state architectural artefacts for the solution.

Artefacts from all phases, including logical design, gap analysis reports and others, are also lodged with enterprise architecture for future reference. These must include any changes to the design that have occurred through the processes of physical design, implementation, deployment and validation.

Other models in the enterprise architecture repository or continuum that have been changed as a result of the implementation of the solution can be validated and signed off as current. What was the target architecture or future state for the solution thus becomes the current state and the basis for business-as-usual operations. This new current state is also the starting point for future solution architecture work.

3.11 BENEFITS AND CHALLENGES OF USING AN ARCHITECTURE FRAMEWORK

Frameworks are useful in multiple situations where standardisation of practice is required. However, unlike a strict methodology where activities are mandated, a framework is meant to be a guide, not a driver or constraint. Instead, frameworks offer suggestions as to what needs to be done and provide options for completing tasks.

Some benefits of using a framework for solution architecture include:

- **Captures best practice:** which may be introduced as part of a framework but is then modified and made to work within the structure and culture of the organisation. The framework can be used to make it pervasive throughout the organisation or enterprise.

- **Promotes consistency:** when doing solution architecture work because, by using a framework, the same techniques and standards that others have found to be successful will be reused. This has the added benefit that practitioners in different parts of the business will understand the common approach and make them more flexible and mobile in terms of working across the organisation or enterprise.

- **Builds confidence:** among stakeholders who see consistent results, common communication tools and know what to expect from solution architecture.

- **Increases reusability:** through consistency of approach and the standard definition of artefacts such as views and viewpoints, work done as part of one architecture project will be beneficial with less rework by another architecture practice, whether in solution or enterprise architecture.

- **Increases maturity:** of solution architecture practice within the organisation as the framework acts as an induction and training tool for new practitioners and a quality assurance tool for the organisation or enterprise.

However, the use of a framework for solution architecture does present some challenges, including:

- **Finding suitable tools:** which can be immensely beneficial as they can assist with automation of tasks and standardisation of documentation, among other things, but careful research and selection is needed and this can take time and resource. A mistake in selecting the correct toolset for managing solution architecture can be costly if it fails to provide the value promised. An ineffective tool can slow down the adoption of solution architecture by the organisation or enterprise.

- **Justifying the costs:** of setting up and implementing solution architecture, which may be a completely unknown discipline, meaning that unworkable limits may be put on investment in personnel, tools and access to stakeholders. For organisations or enterprises where architecture is completely new, the best approach may be to start small and achieve some success before requesting substantial investment.

REVIEW QUESTIONS

1. Which phase in the solution architecture life cycle has the gap analysis report as one of its outputs?

   a. Roadmap development.
   b. Discovery.
   c. Validation.
   d. Solution outline definition.

2. What term is used to describe one of the criteria that the solution must fulfil in order to be successful?

   a. Governance.
   b. Viewpoint.
   c. Stakeholder.
   d. Requirement.

3. Why is it beneficial to use visual representations during solution architecture? (Select the **best** answer)

   a. They are quicker and easier to produce.
   b. They are more easily understood and good for communication.
   c. They are preferred by architects who always think visually.
   d. There are more tools that support visual diagrams than text-based artefacts.

4. Which of the following elements are included with each option in a business case?

   i.   Interface analysis.

   ii. Cost benefit analysis.

   iii. Gap analysis.

   iv. Trade-off analysis.

   v.   Description.

   a. ii, iii and v only.
   b. i, ii and iv only.
   c. ii and iii only.
   d. iv and v only.

4 INPUTS TO SOLUTION ARCHITECTURE

LEARNING OUTCOMES

When you have completed this chapter, you should be able to demonstrate an understanding of the following:

- How internal drivers for change generate the requirement for a solution
- The impact on solution architecture of external drivers for change
- Technical requirements, standards, principles and policies
- General business requirements, objectives and corporate governance
- Functional solution and component requirements
- NFRs and quality of service (QoS)
- The types of baseline artefact that can be selected as inputs
- Factors that constrain solution design

4.1 INTERNAL DRIVERS FOR CHANGE

Although solution architecture is delivered within the context of a business vision, it has its own process with a series of milestones, deliverables and dependencies. In supporting the business to achieve its objectives, all design decisions must be driven by the business. This applies from the start at the initiation phase, all the way through the life cycle to the point at which a working solution is delivered and approved.

At the beginning of the solution architecture life cycle, critical information is handed over from the business to the solution architecture team, consisting of:

- **The problem statement:** contained in the solution vision statement.
- **Any other useful documentation:** this varies enormously depending on the origin of the idea for the solution.
- **Authorisation to find out more:** contained in the architecture initiation document.

The solution vision statement and the architecture initiation document are formally defined artefacts that contain enough to go ahead to the discovery phase. These two artefacts, together with any other useful documentation that may exist, constitute the initial inputs to solution architecture.

During the discovery phase, more inputs are obtained through investigation by the solution architecture team, working with stakeholders from the business. Inputs will also be rationalised and elaborated during this phase to make them as relevant as possible to the solution being designed.

4.1.1 Business strategy

An organisation or enterprise's **business strategy** sets the **direction** and **priorities** for the business operation, and identifies what is needed to make progress towards its goals. It embodies the business's motivation for improvement and development, describing why things need to change as well as what aspects of the business need to be the focus of action.

The business strategy is a very important source of ideas for change, some of which may become solutions and require design by solution architecture. In other words, the business strategy articulates problems that are seen as a priority for action. If the action is to ask solution architecture to investigate and design a solution, then the problem must be documented in the solution vision statement during the initiation phase of the solution architecture life cycle, which becomes an input to solution architecture.

Apart from the problem that has been identified, some additional concepts from the business strategy may be relevant to the solution and therefore useful as inputs to solution architecture. These include:

- business requirements;
- business objectives and targets;
- the results of prior investigation and analysis;
- artefacts from the business strategy such as plans, maps, models, charts and value chains.

The business strategy can also be a source of constraints, for example where existing plans overlap with the problem under consideration.

During the discovery phase, the business strategy is used as a reference point to provide assurance that decisions are being made in line with the needs of the business.

4.1.2 IT strategy

Any IT strategy must be inextricably bound to the business strategy, since the purpose of IT is normally to support the business aims and activities of the organisation or enterprise. If there exists an active infrastructure or technology function, then it is responsible for implementing the IT strategy, as well as playing a major role in its formulation and definition.

Like the business strategy, the IT strategy does not start with a blank canvas but is always working from the current situation to a desired future state, based on a set of higher-level objectives and principles.

Not all ideas for change originate in the business, though. New technology is constantly becoming available, and existing technology is constantly developing and evolving. These changes, which are entirely separate from the main focus of the business, can present opportunities and challenges, with the potential for significant impact on business operations.

Fallowdale Hospital patient communications: IT strategy

The following technology developments have led to the IT strategy being updated:

- Availability of cloud-based IT that has a lower running cost and improved performance.
- Ability to provide web and mobile user interfaces for most existing and all new applications.
- Availability of fast broadband, enabling some staff to work remotely, all or part of the time.

Having become aware of these changes following the appointment of a new IT director, the Fallowdale Hospital IT strategy now aims to take advantage of the above developments wherever possible.

Note that some additional infrastructure has already been identified as a requirement to guarantee the necessary connectivity involved in these new developments and to the security infrastructure and policy to take into account the additional risks that have been identified.

There are also challenges from the legacy of existing technology. Part of nearly every IT strategy is concerned with rationalising the infrastructure estate towards using preferred standards, systems and components. Many organisations use a banded **technology radar** (see Figure 7.2) system of ratings to specify an optimal set of technologies (somewhere between leading-edge and legacy) that are preferred for use in new solutions. Others categorise technologies as red, amber or green, with the latter category containing the preferred options.

Every new solution offers the opportunity to take advantage of new developments in IT. Decisions documented in the IT strategy may become technical requirements for a new solution.

IT strategy can also identify technical constraints on solution design. For example, where a new technology or infrastructure component is being deployed as part of a wider rationalisation of the estate, the timing of the deployment could constrain the implementation and deployment of a solution.

4.1.3 Business analysis

Business analysis is an important and widely used discipline and many business analysts and business architects use a five-stage **life cycle for business change**. The five stages are known as align, define, design, implement and realise. The **align** stage examines the external and internal business environments. It identifies ways that the business can become more aligned externally, with its market and the wider world, and with its internal processes and systems. It is at this stage that ideas for change occur, particularly where they are motivated by the need to align the business systems, including IT, with the business strategy.

The external focus in the align stage, which seeks information to help the business to align with the external environment, uses techniques such as PESTLE and Porter's five forces to find areas, opportunities and threats from outside the organisation or enterprise.

The **align** stage of the life cycle is a source of potential requirements for a solution as it may identify problems, risks and opportunities. The **define** and **design** stages are the main focus of the work of solution architecture. The **implement** and **realise** stages are mostly the responsibility of P3M and delivery teams, but solution architecture has responsibility for governance of the solution's delivery. The life cycle for business change and its overlap with solution architecture are illustrated in Figure 4.1.

Figure 4.1 The overlap between the life cycle for business change and solution architecture

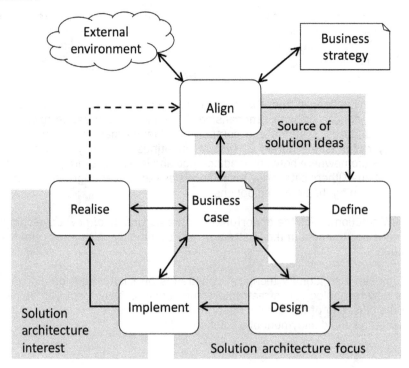

Apart from identifying the need for a solution, the align stage may also provide useful inputs to solution architecture that were obtained during the investigation, including:

- business process models;
- requirements;
- stakeholder details and opinions;
- illustrations of the problem area such as rich pictures, fishbone diagrams and mind maps.

The 'define' stage of the life cycle for business change identifies business requirements and constraints for any changes that will be implemented, as well as articulating the benefits. This overlaps with the discovery phase of solution architecture where inputs such as business requirements and stakeholders are captured and documented. Business analysis can provide support throughout the solution architecture process and business analysts often become members of the solution architecture team.

4.1.4 Enterprise architecture

Enterprise architecture and solution architecture work in harmony to ensure the solutions in use are the best possible fit for what the enterprise needs at any given point in time. Both levels of architecture work on the assumption that the enterprise will endure and therefore design for the long term. Because of volatility and uncertainty, businesses will only survive by being flexible and having the ability to change quickly in response to external factors. Enterprise architecture ensures that solutions both address the immediate needs of the business and are taking it forward in a coherent and strategic direction.

The higher-level strategic view of enterprise architecture can identify problems and opportunities that are candidates for solution architecture.

There are a number of enterprise architecture frameworks in use at present. Some are designed to operate in a **specific domain**, such as the US military's **DoDAF**, while others are **generic**, such as **Zachman** and **TOGAF**.

The Zachman framework, for example, consists of a six-by-six grid with the rows representing stakeholder perspectives and the columns representing architectural elements. This is very useful in ensuring that all elements are looked at from multiple perspectives and that nothing is omitted. It acts as a sort of two-dimensional checklist.

Each column is headed with a generic question that represents an aspect of the enterprise that can be modelled and captured within artefacts and architecture descriptions:

- **What:** examples include details of products and services and other information and data that are significant for the enterprise.
- **How:** methods and techniques used to achieve outcomes.
- **Where:** locations and geographical aspects of the business.
- **Who:** people and organisational structures.

- **When:** timing and life cycles of events.
- **Why:** motivation and purpose.

Each row is labelled with a stakeholder perspective, from the more strategic at the top to those involved in the execution of the strategy at the bottom. The labels vary depending on the situation in which the framework is being used, but some typical labels are:

- **Strategist:** the owner of the business.
- **Executive leader:** directors and senior managers.
- **Architect:** people charged with high-level design.
- **Engineer:** people charged with detailed design.
- **Technologist:** people charged with implementation.
- **Worker:** those who deliver business services.

When used for enterprise architecture, the Zachman framework views the whole organisation or enterprise, seeking gaps or problem areas that can be a source of ideas for new solutions. The Zachman framework can also be used to investigate a specific problem and collect together artefacts and supporting information that are useful as inputs to the solution architecture life cycle.

The TOGAF standard from the Open Group is an enterprise architecture framework and methodology that contains a large amount of detail on the process and artefacts used in the production and delivery of change at the enterprise or organisational level. Since many of the benefits of solution architecture depend on the broad strategic perspective and organisation-wide nature of enterprise architecture, TOGAF or something like it is invaluable to ensure that the necessary artefacts are available for a solution architect to use as inputs.

The TOGAF methodology has a life cycle for the production of architecture descriptions called the architecture development method, or ADM (TOGAF 9.2, 2018). This consists of a number of phases that govern activities such as producing artefacts, recording decisions and interacting with stakeholders:

- **Phase A:** Architecture vision.
- **Phase B:** Business architecture.
- **Phase C:** Information systems architectures.
- **Phase D:** Technology architecture.
- **Phase E:** Opportunities and solutions.
- **Phase F:** Migration planning.
- **Phase G:** Implementation governance.
- **Phase H:** Architecture change management.

Phases B (business architecture) to D (technology architecture) concentrate on describing the current state and a possible desired future state of various domains within enterprise architecture, beginning with business, then working through data and applications and ending with technology. Looking forward to a future state or target architecture in each of these domains opens up the possibility of identifying and prioritising some changes to one or more of these domains that might become solutions. Each of these phases includes a gap analysis to identify what would have to change to achieve the target architecture. This gives a good indication of the type and scale of work that may eventually form part of a project plan or roadmap.

In phase E (opportunities and solutions), the results of the work done in phases B to D are analysed to look for any opportunities and solutions that the organisation is likely to wish to progress. Potential SBBs, and their related architecture building blocks (ABBs), are itemised and rationalised into a single matrix.

Phase E drills down to examine the impact across the organisation including:

- risks;
- interoperability and dependency between solutions and their component parts;
- competing requirements from other parts of the business.

This is only a brief summary of the activities carried out in these four phases of the ADM, but nevertheless its relevance to solution architecture can be seen fairly clearly. Solution architecture aims for a holistic solution to a business problem requiring the high-level view of the organisation's structure and behaviour, which is delivered in phases B to D. The involvement of a solution architecture team begins in phase E, where potential solutions are identified and the work of designing, specifying and delivering their components or building blocks is initiated.

4.2 EXTERNAL DRIVERS FOR CHANGE

A driver is a force that means the organisation or enterprise feels pressure to make changes to part of the business, which could be achieved with a new or modified solution. Some drivers have unambiguously external causes, such as changes to legislation or the actions of competitors. Others are more nuanced, such as strategic decisions by business leaders within the organisation or enterprise that are, in part, responding to external events. No business exists in a vacuum, so it could be argued that all the actions taken by a business are in some way influenced by external factors.

Some drivers for change that have strong external factors include:

- Finance, such as the availability of capital, terms of contracts and relationship with providers such as banks and government.
- Shareholders and other governing bodies that may request changes.
- Regulatory requirements and changes to regimes.
- Competitors and other actors in the same business environment or domain.
- Customer feedback.
- New and modified legislation.

These drivers may already be issues for the organisation or enterprise, in which case the response will be more reactive than proactive. There are some useful techniques that can allow the business to be more proactive by:

- anticipating external change;
- determining the impact;
- deciding on possible actions that can be taken.

If the actions taken are to reduce the likelihood or impact of risks to the business or to take advantage of opportunities, then they are more proactive than reactive. Some techniques that support proactive change are described in the following sections.

4.2.1 PESTLE analysis

This is a technique that aims to capture potential external drivers by dividing the business environment into six categories that form the acronym PESTLE:

- **P**olitical
- **E**conomic
- **S**ociocultural
- **T**echnological
- **L**egal
- **E**nvironmental

PESTLE analysis is a way of stimulating thought. It invites people to come up with ideas by asking the same question from six different points of view: 'Is there something _____ that could affect our business?' where the blank can be filled with one of the adjectives in the PESTLE acronym.

The technique can be used in a broad strategic way to try to gain insights across the whole business or can be more narrowly focused. For example, PESTLE analysis can be used to find inputs to solution architecture to ensure that no external factors that are relevant to the problem area have been overlooked.

Fallowdale Hospital patient communications: PESTLE analysis

- **Political:** government policy, such as the distributed model of care to be nearer to the patient, will lead to more patient communication than the current centralised model.

- **Economic:** pressure on budgets is constant, but the economic situation is unlikely to mean more spending on healthcare by governments or individuals. Any change to patient communications must aim to reduce cost or be cost-neutral.

- **Sociocultural:** patients expect and respond better to instant methods of communication whilst printed letters are often ignored. Patients demand better information, especially on healthcare. There is a high and increasing level of medical misinformation of several types, which affects attitudes to healthcare.

- **Technological:** over 80 per cent of adults in our catchment area have smartphones and of those, over 90 per cent use them as their primary means of communication. Multiple methods of communication are used, including messaging and social media apps. Encryption and identity management technologies are available on smartphones but vary by device, operating system and app. Over 95 per cent of our patients or their carers have access to some form of electronic communication.

- **Legal:** there are new legal requirements for tracking communications. The data protection laws are currently being amended and strengthened. Equality of access to healthcare is embedded in law.

- **Environmental:** moving away from printed communications reduces the hospital's carbon footprint.

4.2.2 Porter's five forces framework

The five forces approach (sometimes abbreviated to 5F) was created by Michael E. Porter (Porter, 1979) and is a method for analysing the external competitive environment of a business. The five forces identified in the framework are:

- **Competitive rivalry:** the strength of competition in the industry.

- **Supplier power:** the ability of suppliers to drive up the prices of inputs.

- **Buyer power:** the strength of customers to drive down prices.

- **Threat of substitution:** the extent to which alternative products and services can be used.

- **Threat of new entry:** the ease with which new competitors can enter the market and drive down prices if they see that good profits are being made.

Porter's five forces analysis is used to analyse the level of competition within an industry or sector, especially when starting a new business, designing a new product or service, or entering a new business area.

4.2.3 Voice of the customer (VoC)

A key external driver for organisations and enterprises of all types is the behaviour of customers. The term **customer** includes past, current and potential future users of the business services provided by the organisation. A customer may be an individual or a business. Customer behaviour includes **engagement** (or otherwise) with the business and **opinions** and communication (e.g. reviews) about the business. The term **business service** includes the supply of physical materials and **products**.

Arguably the customer is the most important source of drivers for change for a business. However, it is not always straightforward to identify the customers of a business. Past and current customers are those who have already used the services provided by the business. Acquiring potential future customers may require changes and additions to the range of business services provided.

> **VoC:** is a process to capture customer expectations, wants and needs, and anything else that helps a business to improve the services it provides (Griffin and Hauser, 1993). The process begins by assuming no knowledge and aims to capture qualitative information rather than quantifying some already known factor.

Once captured, the information must be transformed into actions that will modify the business services to improve customer satisfaction. This is described as visualising the gap between expectations and reality.

VoC can be captured in a variety of ways and it is usual to use a mix of these to ensure that nothing is overlooked:

- interview;
- survey;
- live chat;
- social media;
- focus group;
- service and product reviews.

VoC is relevant for solution architecture as a source of new ideas and problem areas that can be addressed by a new solution or modification to an existing one. Because this process does not assume any prior knowledge about customers, it can expose or highlight areas that have not even been considered by the business and are therefore not addressed in the business strategy.

This approach can also help when an idea or problem area has already been identified. Customers can be invited to comment on the problem area and their responses can be used as a source of requirements for the solution that will be designed and developed.

Activity 4.1

Could the VoC process be used to provide useful inputs to the Fallowdale Hospital patient communications solution architecture life cycle?

One specific problem is patients not attending appointments that have been booked for them. Could VoC be used to throw any light on this specific problem area?

What methods of capturing customer views are most appropriate in this case?

What groups of people should be targeted by the hospital in the VoC process?

4.3 TECHNICAL REQUIREMENTS

Technical requirements are sometimes considered constraints as they can be seen as limiting the options available. Seen in this light, they are perhaps closer to NFRs in that they are generally put in place to stop bad things happening or to set minimum standards. Technical requirements may be specific to a solution, but are much more likely to be adopted more broadly across the organisation or enterprise and be applied to multiple solutions as well as in other domains. This makes them the concern of enterprise architecture as well as solution architecture.

Technical requirements, like general business requirements that are covered in the next section, may come from business and IT strategy, corporate governance or enterprise architecture (see Figure 4.2).

Figure 4.2 Sources of requirements and constraints

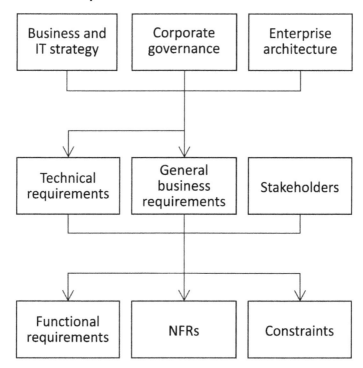

4.3.1 Technical standards

Technical standards are very useful for solution architecture, and for architecture in general, as they capture and embody the **expertise** and **experience** of **professionals** practising over many years and in many situations. Technical standards then present this knowledge in the form of a specification of **processes** to follow and **documentation** to produce, as well as business rules and parameters that must be adhered to.

The generalised nature of technical standards makes them comprehensive enough to work in every situation that the standards committee is able to imagine. This means the standards are likely to contain specifications that do not apply to a given situation or solution. So, whilst it is useful to identify a particular standard that is applicable to a solution, more work is required to clarify which parts of the standard are to be used as part of a technical requirement.

The enterprise architecture repository or continuum contains a catalogue of technical standards that have been used previously within the organisation or enterprise and is a good starting point for finding those that apply to a solution.

4.3.2 Principles and policies

Principles and policies are organisation or enterprise directives that apply to all solutions, so it is not necessary to state in a requirement that a solution must comply with a list of principles and/or policies. Even directives that originate from outside the organisation or enterprise and have been adopted by it must be followed.

Solution architecture can use principles and policies as the source of requirements and make them specific to the solution by first creating a list of directives that are in scope for the solution and then selecting specific parts or aspects of each directive that require action, either during design or implementation of the solution.

Some requirements that have been derived from principles and policies fall within another category, such as functional requirement or NFR. If so, they should be added to the requirement catalogue under this categorisation and linked back to the technical requirement from which they were derived.

4.4 GENERAL BUSINESS REQUIREMENTS

This is a category that contains high-level requirements driven by:

- Owners and business leaders directing the organisation or enterprise through strategic decisions.
- Legislation from the jurisdictions in which the organisation or enterprise operates.
- Agreements with regulatory authorities, industry bodies and business partners.

As with technical requirements that are based on principles and policies, general requirements apply to the whole business, including all solutions. Therefore they need to be made specific to a solution by highlighting the part or aspect that is applicable. Usually this leads to the creation of more specific functional requirements and NFRs. These are linked to the general requirement from which they were derived using cross-referencing in the requirements catalogue, which helps to understand the original motivation and justification.

4.4.1 Business objectives

Business objectives are **aspirational targets** that have been identified by business leaders as instrumental in achieving the **vision** and **mission** of the business. They may be defined using the **SMART model** to ensure they are specific, measurable, actionable, realistic and time-bound. Once an objective has been identified, designing, implementing and deploying a solution is one way of realising it for the business.

A single business objective may be the entire rationale for a solution, but this is not always a one-to-one match; one solution may address multiple business objectives or only make a partial contribution to one. Business objectives may additionally lead to constraints on the design and implementation of a solution. For example, it may be possible to achieve an objective with a well-designed solution, but it will be unacceptable if this hinders the achievement of another business objective in some way.

The most common pattern is that a single business objective is expressed through multiple functional requirements and NFRs. These can be linked to their originating business objective by references recorded in the requirements catalogue.

4.4.2 Compliance with legislation, standards and policies

Legislation and standards are external drivers for change that may impact on a solution by acting as either requirements or design constraints. Policies may originate from within the organisation or enterprise or may have been adopted from an external source such as an industry body.

The requirement to comply with a piece of legislation, a standard or policy needs to be explained in the context of the solution; for example, it is essential to know why it applies to the solution, and which parts are relevant. This may be made more specific by deriving functional requirements and NFRs for the solution and linking them back to the general business requirement by cross-referencing in the requirements catalogue.

4.5 FUNCTIONAL REQUIREMENTS

A functional requirement is a statement of what a solution must do to satisfy the needs of its stakeholders. This type of requirement is distinguished from the inelegantly named NFR category, which specifies the size, capacity, accessibility or other aspect of all or part of the solution. Functional requirements and NFRs can apply to all or part of a solution, such as a component or group of components working together to provide a service. This includes any component of a solution: people, organisational units, business processes, information and technology, as well as IT systems.

For example, an organisation has decided to outsource the transport of items between its business locations. It is a functional requirement to collect an item from a specified business location and deliver it to another specified location when requested to do so.

There are many NFRs that could apply to the transport outsourcing solution. One example would be to specify the number of transport requests that can be made in a specified time period. It would also be useful to specify the maximum time between request and collection, and collection and delivery.

Functional requirements can be placed in categories, including:

- information and data;
- activities from business processes;
- enforcement of business rules;
- reporting, management information (MI), BI and data analytics;
- legislation and regulation;
- transaction processing;
- administration and access management.

It is useful to have such categories for a number of reasons. Categories can help to make sure requirements are not missed or overlooked by asking for requirements in each category. For example, when interviewing stakeholders it is worth asking about each category to stimulate ideas.

Categorisation can also be useful for managing requirements, especially where there are many requirements for a solution. Categorisation allows requirements to be grouped together and handled in smaller batches. Similar, overlapping or duplicate requirements are likely to be found in the same category, allowing them to be merged, made more specific or linked to other requirements.

4.5.1 Information and data

Organisations and enterprises run on information and data, which are integral to most activities. Therefore most of the functional requirements of a solution have an information or data aspect that needs to be captured as an input to the solution architecture process.

Information and data requirements can be divided into four basic categories with the acronym CRUD:

- **Create:** definition, collection and acquisition.
- **Read:** use in a business process, use for decision making, manipulate, analyse, report, alert and transfer to a third party.
- **Update:** respond to event, version control, status change and correction.
- **Delete:** archive and destroy.

Apart from the category, there are a number of other details of information and data requirements that should be captured and recorded. A useful place to record these details is in an information or **data dictionary**, especially for new information or data items that are not currently in use within the organisation or enterprise.

Enterprise data architecture can advise if information or data is currently in use and this is recorded in the enterprise information or data model and in the enterprise data dictionary. All requirements should be logged in the requirements catalogue and given a unique ID for reference purposes.

Some of the details that need to be captured for information or data requirements are:

- business term;
- definition;
- owner;
- requirement ID;
- component ID;
- use;
- constraints.

Fallowdale Hospital patient communications: functional requirements

Information and data requirements for message contents:

- **Business term:** message, currently known as letter.
- **Definition:** a message is a communication from the hospital to an external recipient (initially a patient, parent of minor, or carer).
- **Owner:** operations director.
- **Use:** communicate information to patients about events such as appointments, results of consultations and diagnostic tests, and other information related to their care and management of their condition.
- **Timing:** appointment message to be sent immediately after an appointment is booked or if cancelled by the hospital. There are limits to how close to the appointment the message can be sent, and this varies by communication type. There is a manual process for late notice cancellations by the hospital. Results should be communicated as soon as they are approved by an appropriate healthcare professional.
- **Data records:** the date and time of sending and the destination address must be recorded for management and auditing purposes. This is a legal requirement.
- **Management:** it must be possible to change the content of messages. Specifically the hospital wants to add messages that affect all visitors to the hospital, such as winter virus alerts to prevent those with symptoms from attending.
- **Equality:** any messages must meet the requirements of equality legislation including easy-read English, other languages, large print, Braille, and support for assistive technology such as screen readers and electronic Braille devices.
- **Patient preferences:** the ability for patients to maintain their own communication preferences will save a lot of administrative time.

4.5.2 Activities from business processes

Solution architecture has two distinct perspectives for business processes.

- **Black box view:** business processes are business components within a solution, treated as encapsulated units with defined interfaces. This is called the black box view because the inner workings are hidden. Business process behaviour is expressed through functional requirements and NFRs.

- **White box view:** business processes are a collection of activities. This is called the white box view because the inner workings are visible. Each activity has functionality and may be subject to functional requirements and NFRs. Each activity may be fully or partially automated using technology.

Fallowdale Hospital patient communications: business process

Figure 4.3 shows the business process for patient referrals.

Figure 4.3 Patient referral business process

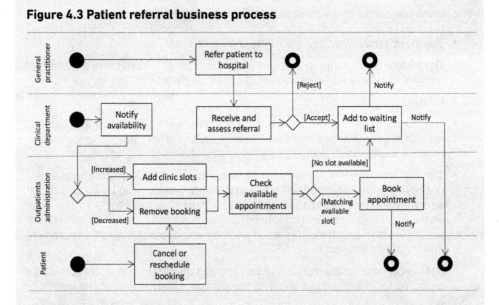

The process has four swim lanes, representing the actors with activities in boxes and decision points as diamonds. Process start points are solid black circles and end points are circles with white centres.

4.5.3 Enforcement of business rules

A business rule is a way of controlling business operations. For example:

- who is authorised to approve payments to external organisations;
- how tasks are allocated to team members;
- when a resource can be released;
- the conditions for escalating to a senior manager.

They can be expressed as a set of **values**, a simple **formula**, or a more complex set of instructions otherwise known as an **algorithm**. These rules are often operated by people and may not even be written down anywhere, for example where only a few people ever apply the rule. This is problematic, especially when new people join the organisation or when scaling up a business operation.

There is also an enterprise architecture perspective on business rules. If a business rule is correct in one part of the business then it might be needed elsewhere. Often business rules are applied or enforced in part of the business but not in others, leading to inconsistent behaviour on the part of staff, customers and even automated systems.

Some business rules may be known about at the initiation phase of solution architecture and may even have been identified as part of the problem. Other business rules may be found during the discovery phase. Business rules are a type of requirement and are fairly easy to document as such. They are often strong candidates for generalisation; that is to say, a business rule that is part of a specific activity may be applicable more widely across the organisation or enterprise if it can be generalised and made more flexible, perhaps with the use of parameters.

Business rules can be a way of automating decision making. There is a software component called a **business rules engine** that can be connected to one or more systems so that a particular rule can be enforced. This can be configured so that one rule is applied in multiple systems. This also helps to promote consistency when the rule changes, as it only needs to be replaced once in the rules engine.

Activity 4.2

The following problems have been identified that relate to the Fallowdale Hospital patient communication solution:

- Some patients do not wish to receive emails from the hospital.
- Some appointment invitations are reaching patients too late for them to make arrangements to attend and some have reported receiving a letter the day before the appointment.
- Only experienced Outpatients staff can allocate patients to appointment slots, as it is so complicated to work out which slots are appropriate.

Can you identify ways in which business rules could help with these problems?

How would any business rules be captured as solution requirements?

4.5.4 Reporting, MI, BI and data analytics

Reporting is a very common requirement to enable managers to do their job by providing a summary of operational activities. During the transition to a new solution, it is often advisable to replicate any management reports, even if they are not required with the new solution. They can provide a fall-back mechanism and help to build confidence with a new way of working. Management reporting is also known as management information and abbreviated to MI.

At a higher strategic level, where the information is used for longer-term directorial decisions rather than for day-to-day management, reporting is known as BI.

Both operational MI and strategic BI reports may be supplied in the traditional tabular format of printed reports (even if they will be delivered electronically) or via a visualisation system or dashboard. Visualisation may be a requirement if it is a key part of operational or strategic decision making, as it can dramatically speed up how quickly information gets to where it can be used in a form that supports the best possible decision process. Where visualisation is required for easy, clear identification of information or trends, it would be a usability NFR.

Reports can also be provided in machine-consumable format for export to other systems or external agencies. There can also be triggers, such as alerts, that cause automated action if any of the values goes outside a specified range. These triggers can be controlled using business rules.

MI and BI specifications are a type of functional requirement and can be captured as such with details of contents, frequency and which people will receive them.

With rapidly changing information and data, where trends are significant and need to be found and acted on, data analytics and even data science can be brought into play. This area is far less predictable and the requirements that can be defined are therefore less specific. For example, it can be stated that data analytic functionality is required, but not what analysis will be performed. Usually the broad areas of information and data that will be subject to analysis can be specified as part of a requirement.

Fallowdale Hospital patient communications: functional requirements

Some requirements that have been identified for reporting, MI, BI and data analytics are:

- **Appointment status:** uninvited, invited, confirmed and chased, including short notice availability.

- **Cost of communication:** per appointment, per episode, by clinic and by department.
- **Appointments missed:** per period, breakdown by clinic and department.
- **Patient preferences:** known, including language(s) and consent given.
- **Reports to regulator**.
- **Reports to government departments and agencies**.

4.5.5 Legislation and regulation

It seems trivial to state that solutions need to comply with the law of the land (or lands) in which they will operate. Equally, who would want to risk fines and other regulatory sanctions by failing to follow the rules of regulators and industry agreements? However, it is not enough to list the laws and regulations that must be obeyed, although that is a good starting point. Rather, each law or set of regulations must be examined in more detail to extract the exact requirements and their specifications that are demanded of a solution.

In some cases a change in legislation or regulation may be the driver for the solution being considered or for a modification to an existing one.

Enterprise architecture can help to list relevant laws and regulations that are in scope for the problem area being addressed by the solution. The enterprise architecture repository or continuum contains a list that applies to the whole organisation or enterprise. This is a good starting point. Any new items that are not in this list but affect the problem or solution in focus must be added to the enterprise list.

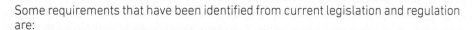

Fallowdale Hospital patient communications: functional requirements

Some requirements that have been identified from current legislation and regulation are:

- **Data protection:** ability for patients to see details of communications that concern them; hospital data archival and destruction business rules must be used.

- **Privacy:** personal information is currently protected to a high level, interfaces and new information must not weaken this protection; patients must be able to consent to the use of data for communications (this is not currently the case with postal communication).

- **Equality:** patient communication must be equally accessible to all patients so more details of language, disability and preferences must be captured and managed; communications must be written in accessible language and use automated checks for this where possible.

- **Incident investigation:** internal governance and external regulation require the ability to access all records, including those about communication, that relate to the care of an individual patient; therefore functionality to allow this in a clear and transparent way must be included in the solution (this currently takes a lot of time with postal communication).

4.5.6 Transaction handling

Transactions are usually short-lived and high-frequency, approximately equivalent to a business process activity. They represent **core business activities** and are therefore important **sources of requirements** for the solution.

Transactions can be identified by using techniques such as **customer journey mapping**, which aims to capture touchpoints such as:

- **Actions:** performed by the customer or enterprise.
- **Events:** that happen independently of any action.
- **Decisions:** choice between two actions.
- **Motivation:** reason for starting, ending or continuing the journey.

Fallowdale Hospital patient communications: functional requirements

Some transactional requirements that have been identified from customer journey mapping are:

- **Outbound communication:** ability to send messages individually or in batches.
- **Inbound communication:** ability for patients to confirm, cancel or reschedule appointments.
- **Events:** when a patient is booked into an appointment slot, can this automatically generate a message? Can this be done singly or in batches once a whole clinic is filled? Can there be a manual override on this automation?

4.5.7 Administration and access management

Any system needs to be administered. Administration goes beyond day-to-day management, whereby business activities are supervised and directed by operational staff, and usually concerns less frequent activities.

For example, the manager of a theatre is concerned with selling tickets and ensuring there are sufficient staff to make performances run smoothly. Occasionally there are administrative tasks such as changing the price structure or adding a new job role.

Access management is concerned with security, in that it allows trusted staff such as managers to grant access to business resources and the authorisation to perform tasks on behalf of the business.

Administration and access management are both concerned with tasks that are less frequently performed than core business activities but are nevertheless essential to business operations. Therefore they are a source of functional requirements.

Fallowdale Hospital patient communications: functional requirements

Some administration and access management requirements include:

- Modify clinic message content.
- Modify hospital message content.
- Add, remove and modify permissions based on job role.
- Set limits and receive audit alerts when exceeded.
- Determine which events are recorded in the audit trail.
- Produce operational reports and alerts (bottlenecks, blockages and failures).
- Handle rejected communication, such as non-existent addresses.

4.6 NFRs

This type of requirement specifies a characteristic of part or all of the solution that is essential to its success in addition to the functionality it provides. NFRs are sometimes known as quality attributes and they can contribute to the QoS of a piece of functionality. Note that the quality of a solution depends on satisfying both functional requirements and NFRs.

If an NFR applies to part of the solution, that part could be a component or group of components that work together to provide some functionality, sometimes known as a subsystem of the solution. NFRs may be directly linked to one or more functional requirements but are more likely to apply to a structural unit such as a component, subsystem or the whole solution.

For example, a functional requirement to provide weekly management reports could have specific NFRs such as the currency of the information (how up to date it is). Management reports, especially periodic ones, are usually more tolerant of non-current information of data because they are retrospective and are used to monitor trends over time.

Quite often, however, information currency NFRs apply to the whole solution because of the interconnectedness of components; in other words, out-of-date information in one part of the solution can manifest elsewhere.

However, requirements are supposed to state and justify what is required and why. It is a mistake (often a costly one) to push component NFRs up to the level of the whole solution 'just in case'.

4.6.1 Performance

Typically performance is measured in terms of speed. For interactive systems this is the response time experienced by the user when interacting with a system, whether that be a slow computer screen that keeps the user in suspense with an hourglass or spinning circle, or the familiar on-hold telephone help desk music. Users consider 0.1 second as instantaneous but can stay focused on a task for 1 second; more than 10 seconds and they will start doing other things (Nielsen, 1993).

For automated processing, a common measure is transactions per second (TPS). The speed of delivery, usually measured by **elapsed time** to complete a task, is a measure of performance that can be used to specify the behaviour of a solution or component using an NFR – for example, how long it takes to send an email once it has been prepared.

Both interactive and automated systems, as well as business processes and other activities, can be specified using a type of average measure known as **throughput**. This specifies how many of a certain task will be performed in a given time interval.

Fallowdale Hospital patient communications: NFRs

Performance NFRs:

Some considerations for performance are:

- Time taken to send a message.
- Number of messages per hour.
- Response time of patient-facing systems such as setting communication preferences.
- Response time of connected systems (organisation and patient administration systems, clinic and waiting list systems).

Note that for existing systems, NFRs will be considered for their impact on the solution in focus and vice versa. Even if the patient communication solution does not require an increased performance from the existing systems, they need to be set and measured in case the new solution degrades the performance of an existing system.

4.6.2 Capacity

A system or business service may have requirements of scale depending on the amount of activity that is anticipated. Increasing the scale to accommodate more activity usually means adding more resources such as technology, people or equipment. Capacity NFRs can have a minimum and maximum level and also specify how quickly the solution component can scale up or down, known as scalability.

Capacity can apply to a solution in three different ways:

- **Activity:** how many instances of an activity can be completed in a given time.
- **Storage:** how much of a resource can be stored.

- **Scalability:** minimum and maximum capacity and how quickly additional capacity can be added.

Fallowdale Hospital patient communications: NFRs

Capacity NFRs:

- Starting capacity is not greater than the current number of letters sent by post.
- Initially, only invitations to outpatient appointments are being sent to patients using the new solution, so this will be less than starting capacity.
- This will grow to include all current communication by post, equal to starting capacity but allowing for 10 per cent growth.
- Thereafter it will include other communications such as diagnostic results and communications with other healthcare professionals. These have been estimated, but more accurate and precise figures are being worked on.

4.6.3 Integrity (consistency and accuracy)

Integrity is an NFR that is usually applied to information and data that needs to be consistent and accurate. It seems intuitively obvious that inconsistent and inaccurate data can cause problems, but it is less obvious how this can be specified in an NFR.

When considering consistency, it is worth thinking about the ideal situation and also about potential problems. Note that most people have experienced the problems, but the ideal state is more elusive. The ideal is that information and data are completely accurate and up to date throughout the solution. Breaking this down further, information and data within the solution need to be complete and accurate in two separate ways:

Internal: information and data are identical between different parts of the solution.

External: information and data are the same as their real-world values.

Integrity and accuracy can be controlled using information and data design and specifying how information and data are captured, processed, stored and transmitted. This design can then be enforced using various technologies such as data storage and schema validation. Information and data can be audited against the design and the results used as a measure of compliance.

Fallowdale Hospital patient communications: NFRs

Integrity NFRs:

- Due to concerns about accuracy in main patient database, any details used for patient communications must be validated and verified within 24 hours before use.
- 100 per cent compliance with data model for message data.
- If time of sending message is greater than 60 seconds, then any event content information must be checked to see if still current.

4.6.4 Availability

Availability of a service or system is defined as the 'property of being accessible and usable upon demand by an authorised entity' (ISO/IEC 27000:2018, 2018). In simple terms, this means being there and working when it is needed for those authorised to use it.

A common specification of availability is known as uptime (when the system or service is available or up), expressed as a percentage. Sometimes availability is measured by downtime (when the system or service is unavailable or down), usually to allow for maintenance. This can be further refined by date or time specifics, such as periods when the functionality is not required or allowances for maintenance on a periodic basis during times of low demand.

Other factors that affect availability and which can be specified as NFRs, usually with maximum allowed values, are:

- **Resilience:** measured by the frequency and severity of faults and failures.

- **Recoverability:** time lost while waiting for resources such as data or people to become available after a failure.

- **Serviceability:** time lost when updating the solution to correct errors or add functionality.

Fallowdale Hospital patient communications: NFRs

Availability NFRs:

- Current system is driven by staff allocating patients to slots and then issuing a batch instruction for letters to be printed. The implied SLA for availability is therefore 09:00 to 17:00, matching the working hours of the Outpatients staff. The letter printer needs to be available from 08:00 to 18:00 as it is used for other tasks. Neither of these components is required to be available outside of these hours. The printer is powered off when not in use. The patient administration system has a stated availability of 24/7/365 or 100 per cent, but in practice it does not meet this and records for the last year show it to be approximately 97 per cent.

- The new system will have different requirements, partially because of patient involvement and partially because of automation. Both require 24/7 operation.

- Availability NFR has been set at 99.5 per cent, with appropriate holding messages displayed when unavailable.

- Resilience NFR allows for failure of system components to occur on average once per quarter.

- Recoverability NFR specifies that system services are restored to full operation within 90 seconds, data loss that breaks transactional integrity must be restored within 360 seconds of detection, catastrophic data loss must be recovered within 30 minutes of detection.

4.6.5 Usability

> **Usability**: the extent to which a product can be used by specified users to achieve specified goals with effectiveness, efficiency and satisfaction in a specified context of use (ISO 9241-11:2018, 2018). Other aspects for consideration include error tolerance, learnability or memorability, and engagement.

Usability seems a difficult concept to pin down and specify, and even harder to measure, but some ways to quantify this aspect of a solution are available:

- **Effectiveness:** users can successfully complete actions. With electronic systems, completion rates can be collected and analysed. Success must include correctness and accuracy, so the number of errors is a negative measure of effectiveness.

- **Efficiency:** users complete actions quickly and with minimum effort. Speed can be measured but effort needs more thought.

- **Engagement:** users understand what they are doing and why. This can only really be measured through user feedback.

- **Error tolerance:** users complete actions with no errors or error messages. Errors can be measured, but error tolerance really means that errors are easily recoverable without external intervention such as a help desk. This relies on detecting the error and giving good-quality, helpful feedback. Calls to a help desk and the reasons for them can also be measured.

- **Ease of learning:** new users are effective and efficient and improve with multiple use. This can be measured by monitoring call centre activity.

Fallowdale Hospital patient communications: NFRs

Usability NFRs for the patient response interface (for selecting, confirming, cancelling and rescheduling appointments):

- Task completion rate > 60%.

- Average task completion time < 2 minutes for confirming and cancelling, < 5 minutes for selecting and rescheduling.

- Digital conversion rate from using phone > 40% in first six months.

4.6.6 Accessibility

Accessibility refers to the design of business services, products or environments so that they can be used by all people irrespective of potential barriers such as disability, language or other constraints that are present in the population being served. The aim is to enable everyone to achieve identical results and gain the full benefits of the service, product or environment.

Accessible design can be realised through direct, unassisted access and by indirect access, meaning compatibility with some form of assistance including assistive technology such as a screen reader. An example of accessible design for direct access is the use of well-known symbols rather than words, as these are understood by most people.

Research and development in accessible solutions can be beneficial to everyone.

Activity 4.3

What aspects of the Fallowdale Hospital patient communications solution require accessible design? Can you distinguish between direct and indirect access for any of these design considerations?

4.6.7 Security

Security features can be specified as functional requirements of a solution where security activities need to be carried out by users or actors who interact with the solution. As an NFR, security can be quite difficult to pin down. However, one area that should be quite clear is access and authorisation control. This is a specification of which people and systems should have the ability to perform tasks and access resources. Conversely, any person or system not in the specification for a task or resource must be prevented from gaining access.

A security specification in its simplest form is a cross-reference grid where permission is indicated at the intersection of a resource or task on one axis, and a person or system on the other. Note that 'person or system' is often generalised to 'entity' and may include groups, roles and other multi-entity constructs.

There are a few additional details that need to be specified for a solution such as the type of access allowed. Read access may be permitted for some entities and write access reserved for only those where it is absolutely essential to complete a task. Tasks themselves may be complex with multiple parts and some entities may be given access to perform only part of a task. There may also be conditions that need to be satisfied for access to be provided.

On the negative, preventative side of security, it is necessary to specify the level of protection that is required to stop unauthorised access to resources and functionality. This can only be done as part of an NFR by giving each resource and task a security level (low, medium, or high, for example) based on an assessment of the severity of any breach. The implementation of such a security specification needs the expertise of security and infrastructure specialists.

Apart from gaining unauthorised access to resources and functionality in the form of tasks, security attacks can target non-functional aspects of a solution such as availability and performance. These are sometimes known as denial-of-service attacks. Again, the implementation of protection measures will be done by specialists in the field, but solution architecture needs to specify the requirement in terms of the seriousness of the impact on the solution were such an attack to be successful. This needs to be done for every NFR.

Some measures that can be put in place as SLAs include:

- number of breaches over time;
- successful/unsuccessful attacks;
- penetration test failures;
- audits of compliance with security specification.

Fallowdale Hospital patient communications: NFRs

Security NFRs include:

- As the email and postal address details are classified as personal information, the minimum data set must be collected and transmitted at every stage to comply with data protection regulations. The solution design must contain a business use justification for any data items in this category.

- A complete audit log of access to the patient communication systems, including existing components associated with letter printing, must be kept for six months following any access.

- Access to the patient communication solution must be identifiable to the individual staff member.

- Any failed attempts to access the patient communication systems and any suspicious access (for example, outside of working hours or large data volumes) must be alerted to a senior staff member as well as to IT security management.

- All electronic data transmission, internal and external, involving patient data must be encrypted using the advanced encryption standard (AES).

- Non-electronic (postal) communication must only be sent to the recipient's verified address.

4.7 BASELINE ARTEFACTS

A **baseline artefact** for a solution is one that represents the state before the implementation and deployment of the solution. This is distinct from a **target artefact** that represents the desired future state.

Some baseline artefacts will be modified as a result of solution design and a corresponding target artefact will be produced. Baseline and target artefacts can be compared using **gap analysis** to identify the changes that need to be made to move from baseline to target. Baseline artefacts are not necessarily modified by solution architecture but are inputs to the process due to the information they contain.

To be useful as an input to the solution architecture process, the artefact must be **in scope** for the solution or have been identified as within its **zone of impact**.

If an artefact is in scope for an application, it will contain existing components and interfaces that may be included unchanged, or require modification, or may need to be removed. The components, their interfaces and the artefacts that model them will be the basis for designing the solution (see Figure 4.4).

Figure 4.4 Sources of baseline artefacts for a solution

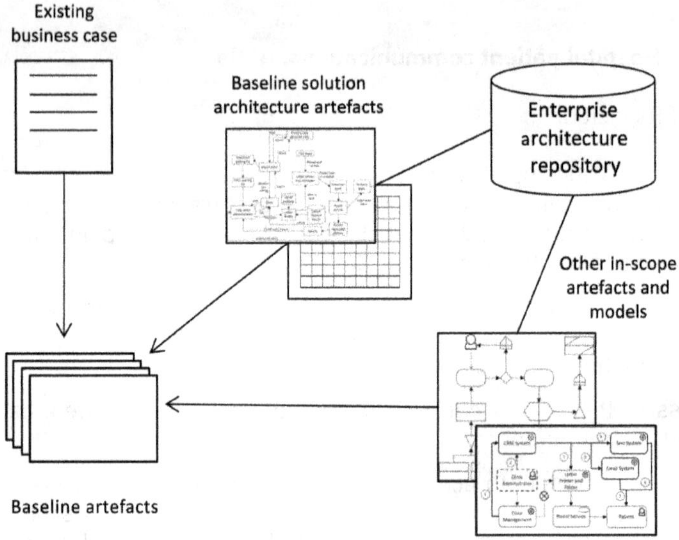

If an artefact has been identified as being within the zone of impact of a solution, for a solution, it will contain components and interfaces that will be impacted by the implementation and deployment of the solution. These impacts will be assessed and modelled during the validation phase of the solution architecture life cycle.

The baseline artefacts that will be modified during the solution architecture life cycle are mainly to be found among those that are in scope for the solution. It is possible that artefacts in the zone of impact may be modified as a result of solution architecture design work. If so, it will be to try to avoid, mitigate or benefit from one or more impacts that have been identified.

4.7.1 Business case

Prior to the initiation phase of solution architecture, a business case may have been created by: a corporate process; P3M; or business analysis.

An existing business case may have been created for several reasons:

- **The proposed solution:** the source of the idea currently being considered, in which case it and its contents are very valuable as inputs.

- **A previously rejected solution:** a related idea that was not taken forward and could provide some valuable insights about stakeholder views and priorities, and so can still be useful as an input to a new solution.

- **The solution in use:** the solution that is being replaced, modified or revised, in which case it serves as the basis for the new business case and as a reference point for what went before.

A business case is modified many times during the solution architecture life cycle. Specifically, it contains details of outline solution design options that have been considered by stakeholders and which of them was selected for detailed analysis and design. The completed business case is lodged in the enterprise architecture repository or continuum and is therefore available for future reference by solution architecture.

4.7.2 Baseline solution architecture artefacts

If a solution is being replaced, improved or updated, then all artefacts for the current solution are critical inputs, especially artefacts that are subject to design. These can be used as the starting point for analysis and design and are also used in gap analysis during the validation phase. The artefacts that are accessible from the enterprise architecture repository or continuum are:

- solution data models;
- component catalogue;
- interface catalogue;
- logical design;
- views and viewpoints;
- gap report;
- gap models;
- impact report.

Note that the baseline solution may not have been designed by solution architecture and may have happened so long ago that there are no artefacts available. In this case, the solution architecture team working on the new solution will have to create design artefacts from scratch. There are usually some sources of the inputs that would usually come from baseline solution architecture artefacts and these should be collected and catalogued. This will have a detrimental effect on the gap analysis to be performed in the validation phase, but enterprise architecture artefacts can often be used as replacements. If there is no existing documentation for an existing solution, it is not usually necessary to recreate it. However, it is important to identify existing artefacts that are in scope for the solution, as these are essential inputs to solution architecture. If such artefacts are changed by the solution, they must be updated in the enterprise architecture repository when the solution is implemented.

4.7.3 Enterprise data models

All data models that are in scope or in the zone of impact for a solution should be obtained and used as inputs. Using data models from enterprise architecture is essential for ensuring consistency of design and will speed up approval by stakeholders, as the data designs are already known and in use within the organisation or enterprise.

Any new information or data that is in scope for the current solution but is either new or a significant extension to the existing enterprise data architecture needs to be tested and approved. By basing any new or extended data models on existing ones, the solution architecture team can ensure consistency and prevent duplication of effort, making sure the absolute minimum is changed to satisfy the needs of the new solution.

4.7.4 Application catalogue

An application catalogue is a central part of the **applications architecture domain** of enterprise architecture. It itemises and gives details of applications within the organisation or enterprise. The details include the **functionality** provided and whether it is **being used**, and in some cases functionality that can be **easily added** by configuration or the acquisition of additional modules. Details of data management capabilities are also documented here. These details are the key to identifying applications that are in scope for the solution.

This is done by examining the problem area and finding applications with:

- functionality currently in use;
- functionality available but unused;
- functionality not available but easily added;
- data currently being managed;
- unused data management capability.

The application catalogue will be updated as a result of solution architecture and design if there are any additions to the application portfolio, modifications to existing applications, or if any are removed from use.

4.7.5 Business models

Business models typically show the **entire business** in a highly abstracted form that gives an overview of business strategy and operations and illustrates the **interconnected nature** of the business components. This makes them extremely useful for gaining an overview of the situation surrounding an identified problem.

Business models can be used to:

- narrow down the scope of a problem;
- highlight what needs to change to solve the problem;
- stimulate ideas about potential options for a solution;
- identify areas of the business that could be impacted by the change brought about by a solution.

Business models are not design documents but can be affected by changes in design. It is possible, but not common, for a business model to be modified as a result of the design and implementation of a single solution. This is not true of other areas in the business architecture domain. Solutions often require changes to business components and these need to be reflected in the artefacts of the business architecture domain. Business models can help to identify a general area of interest within the business. More detailed information about the components and the artefacts contained can then be obtained from the enterprise architecture repository or continuum.

4.7.6 Business service catalogue

The business service catalogue lists all services that are provided by the business to customers, both external to the organisation or enterprise and internal, where this interaction model is in use. Each business service is linked to the business processes that support its delivery and so business services can be linked together where they share a business process. Since business services are the most visible aspect of a business, since they are what the customer sees and experiences, they are often the source of business problems that need to be addressed by solution architecture. For example, customer feedback is usually associated with one or more business services.

The traceability of business service to business process can be continued to other components of the business such as job roles and organisation units. Further details of, for example, data and applications are also easily traced using catalogues and cross-reference grids from the data and applications architecture domains.

Business services catalogues are not design documents but may be modified as the result of solution design, where business services are impacted by a solution.

Fallowdale Hospital business services catalogue

The following have been extracted from the business services catalogue and do not represent the complete list:

- Medical screening (for breast cancer, bowel cancer and other conditions).
- Emergency department.
- Referral for treatment (67 medical departments).
- Diagnostics.
- Maternity.
- Clinical trials.
- Palliative care.

Each of these services requires patient communication and communication with healthcare professional outside the hospital. They are therefore candidates for inclusion in the patient communications solution in the future.

4.7.7 Infrastructure models

Technology infrastructure models are not usually subject to design by solution architecture but rather by **infrastructure or technical architecture**. Infrastructure design responds to the **needs of the business** by providing **infrastructure services** that satisfy the needs of application, data and solution architecture components. For solution architecture, this is done through a process called **solution technology definition**.

Baseline infrastructure models show what infrastructure services are currently being provided, and which ones are being consumed by applications and solutions that are currently in use. If the solution in focus is a replacement or modification for an existing one, the solution technology definition documentation helps to understand the decisions that were made about the realisation of the solution through infrastructure services when it was being implemented.

Infrastructure models can provide a useful insight into some of the physical aspects of the problem domain. Some of these aspects can also help identify constraints on the design of a new solution.

Activity 4.4

The Fallowdale Hospital patient communications solution requires various preferences of recipients (patients, parents and guardians of minors, carers) to be recorded so the right type of communication is used. What baseline artefacts could be useful for this aspect of design and why?

4.8 CONSTRAINTS ON SOLUTION DESIGN

Constraints are limits on design. These often exist as a result of corporate decision making, such as those that derive from the strategy of the organisation or enterprise. These include directives that have been created internally to control the actions and behaviour of all people and external directives to which the organisation or enterprise is compelled or agrees to comply.

Directives include:

- principles;
- policies;
- legislation;
- regulation;
- standards.

Note that these factors can also be the source of functional requirements and NFRs, as well as being constraints. Apart from these universal constraints on solution design, there are a number that vary from one solution to another and these need to be considered on a case-by-case basis.

4.8.1 Time

Time can be a constraint for a solution in at least three ways:

- **Deadline:** all or part of the solution must be in place by a certain time. This could be an external event such as a new law being enacted or a calendar event such as financial year end or the start of school holidays. Internal deadlines also apply, especially where budgets apply, or a commitment to solve a problem by a certain time has been given by business leaders.

- **Dependency:** the solution cannot be implemented or deployed as something is not ready. This could be a component of the solution being designed, or another solution if they are both part of a larger programme of business change with interdependencies.

- **Urgency:** if the problem is hurting the business badly, then the pressure to find a solution or an intermediate solution in a short period of time may override other concerns.

Solution architecture is able to deal with time constraints throughout the life cycle, including:

- Setting limits for architecture work in the architecture initiation document.
- Selecting options in the business case based on gap analysis estimates.
- Selecting components and interfaces based on amount of change required.
- Using intermediate logical designs to deliver benefits in stages.
- Prioritising delivery of certain benefits in the roadmap development phase.

4.8.2 Environment (internal and external)

Internal and external environmental factors may affect design choices. A particular design may satisfy the requirements but be unacceptable because of the internal culture of the organisation or an external sociocultural preference among customers. The factors that should be considered fall into the categories of frameworks such as that of **PESTLE analysis**. This can also be used as a source of drivers for change and may have identified the problem that is being tackled by solution architecture. It is well worth considering environmental factors in the context of constraints, as it would be unacceptable to spend time and effort solving one problem with environmental origins whilst causing another.

4.8.3 Geography

If a solution is to work in a wide geographical area, there are a number of potential constraints that need to be taken into account in solution design:

- **Physical geography:** climate, terrain and distance between nodes.
- **Transport:** infrastructure such as road, rail, airports and sea ports.
- **Resources:** power, raw materials and skilled staff.
- **Human geography:** culture, politics and legal jurisdiction.

It is worth noting that a solution that must operate in multiple geographical locations effectively has multiple environments, so environmental analysis using something like PESTLE analysis should be repeated for each separate location.

4.8.4 Finance

Every organisation or enterprise has some financial constraints on the activities it carries out and solution architecture is no exception. For example, there may be a limit set by the business on the overall cost of a solution.

This can be evaluated as part of the solution architecture design process, in which case the most appropriate way to do this is during business case option selection. Stakeholders get relevant financial information from the cost benefit analysis that is prepared for each option under consideration. There is also a return on investment (ROI) assessment that shows when the financial benefits will offset the costs.

These pieces of information are used to select an option that meets the financial criteria of all stakeholders and specifically the business sponsor or product owner.

The ROI assessment may be critical if finance is tight. For example, it is possible to sequence the design of several intermediate solutions each of which delivers benefits in a short time frame, thus reducing the financial strain whilst still achieving the benefits.

Financial constraints can also be applied at the start of the solution architecture life cycle by limiting the cost of the discovery phase, especially when there is uncertainty about whether a satisfactory solution to the problem can be found.

4.8.5 Capability

Capability may not be a constraint but rather a requirement. For example, business capability is often one of the deliverables of a solution; deficiency in a particular capability, or a shortfall in capacity or competency, may have been one of the drivers for the solution architecture process.

However, there are several areas where capability is a constraint for solution architecture, including:

- **Architecture:** if an organisation or enterprise has a mature practice in enterprise and solution architecture, then this capability enables the consideration of complex solutions. If maturity is lacking, then simpler solutions need to be considered and used to build up the practice of solution architecture through smaller-scale successes.

- **Business change:** organisations and enterprises vary in their capacity for business change. This can be hard to measure and is usually based on past experience by making an assessment of successful and unsuccessful attempts and the lessons learned from them. This will affect the solution architecture life cycle predominantly during business case option selection when the scale and difficulty of achieving change is estimated for each option.

- **P3M:** the delivery of a solution of any significant size is highly dependent on the skills of project, portfolio and programme management. With maturity and experience, complex programmes and projects of work can be delivered successfully and are likely to be high quality in terms of achieving business requirements. Without a strong P3M practice, complex solutions may need to be broken down into more achievable intermediate stages for delivery. This will be reflected in the solution delivery roadmap.

- **Solution development:** during the completion phase of the solution architecture life cycle, one or more teams of solution developers are required to build, test and deploy solution components. This includes software development, business analysis, testing and other capabilities. The type and scale of solution that can be designed depends very much on the capability of those who will build its components.

All of these capabilities are additionally affected by the available capacity and competency of practitioners. In some cases, external capability can be brought in to strengthen internal areas that are lacking. Note that this adds cost and risk to the design and delivery of a successful solution.

4.8.6 Technology

Technological constraints are mainly dealt with by the **infrastructure architecture** team. This may include preferences for certain technological components over others for a variety of reasons that may not be of particular concern to solution architecture. Following logical design, a process called **solution technology definition** determines the technological components that will be used for the solution.

Earlier in the life cycle, technological constraints may become apparent at the solution outline definition phase and when defining business case options for selection by stakeholders.

Technology may become a time constraint if there is a delay in the availability of components that are preferred for a solution design and may lead to alternative technology being used, sometimes as part of an intermediate architecture design.

4.8.7 Risk

Organisations and enterprises have different appetites for risk, some being extremely averse. This should come into play when the options for consideration as part of the business case are being prepared, because each option should be accompanied by a risk assessment. The business case also contains an overall risk assessment of the problem domain. Any risk analysis for a solution should be done in coordination with ERM and appropriate risk policies followed and registers consulted.

REVIEW QUESTIONS

1. An organisation that transports goods wants a solution to provide information for each journey made, including the time taken, average speed, and the number and causes of any incidents that caused delay. What type of requirement is this for the solution?

   a. Technical.

   b. Functional.

   c. Non-functional.

   d. General.

2. A travel company is losing money due to a large number of customer complaints, some of which lead to fines by the regulator whilst others can be resolved with a full or partial refund. The company has a poor record of business change. What types of constraint will be placed on the solution architecture team that has been asked to solve the problem?

   i. Time.

   ii. Environment.

   iii. Geography.

   iv. Finance.

   v. Capability.

   a. ii, iii and v only.

   b. i, ii and iv only.

   c. i and iv and v only.

   d. iii, iv and v only.

3. In PESTLE analysis, what does the 'S' refer to?

   a. Special purpose.

   b. Single.

   c. Stakeholder.

   d. Sociocultural.

4. Which baseline artefact that is used as an input to solution architecture is **least** likely to change as a result of the design and implementation of a new solution?

   a. Business model.

   b. Data model.

   c. Infrastructure model.

   d. Gap model.

5 GAP ANALYSIS

LEARNING OUTCOMES

When you have completed this chapter, you should be able to demonstrate an understanding of the following:

- The purpose of gap analysis
- When gap analysis is used in the solution architecture life cycle
- What inputs are required for gap analysis
- The steps that are carried out
- What outputs are produced from gap analysis and who is the audience for them
- How gap analysis feeds into roadmap development

5.1 THE PURPOSE OF GAP ANALYSIS

Gap analysis is an important tool in enterprise and solution architecture as it is a systematic way to compare two **architecture descriptions** and identify the differences. An architecture description documents an architecture in a particular state, so the differences identified using gap analysis represent what must change to move from the current or baseline state to the target state.

Gap analysis is always performed on **two** architecture descriptions. If there are more states involved in a business change, then gap analysis may be performed multiple times.

For example, where an organisation or enterprise wishes to make major changes to its business, there may be good reasons for breaking this down into stages. One approach is to design multiple intermediate or transition architectures, each of which has an architecture description. Gap analysis can be performed for each transition architecture and the final or target architecture (see Figure 5.1).

Figure 5.1 Multiple gap analyses performed on pairs of successive transition architectures

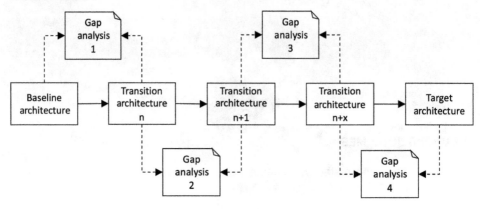

 Note that the term 'gap analysis' has many meanings and interpretations outside of architecture. For example, it is a synonym for 'variance analysis' in accounting and project management, where it is used to highlight areas of concern rather than action.

5.1.1 AD

AD: a 'work product used to express an architecture' (ISO/IEC/IEEE 42010:2011, 2011), the architecture being of a system, such as an enterprise or solution.

An AD represents one possible architecture for the system, such as that of the current or baseline state. Another AD would be required to document the future transition or target architectures. Two ADs would therefore be required to represent the change from the baseline to a target architecture and enable a gap analysis to be performed. An AD may take the form of a set of documents and models or may consist of a set of items in a repository.

An AD contains subunits that are used throughout architecture of all kinds, including **artefacts** and **entities** that are collected together into **deliverables** (see Figure 5.2).

Figure 5.2 AD subunits

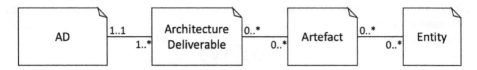

AD: represents the architecture of a system of interest in a particular state. Examples of systems include an organisation, application, business service and solution. An AD comprises at least one architecture deliverable.

Architecture deliverable: a composite document or report produced by an architecture team following an architecture request or as part of an architecture contract or life cycle. Each architecture deliverable is produced for a specific AD. It may contain structured information, in the form of one or more artefacts, and unstructured information, in the form of commentary or analysis. Examples include a statement of requirements, business case, solution outline, logical design and impact analysis report.

Artefact: a catalogue, map, table, diagram or model that focuses on a specific aspect of the architecture of a system, showing the relationship between entities or other components of the architecture. Views are a special type of artefact that are produced multiple times based on a viewpoint (a template for the view) and are linked to one or more stakeholder concerns. Each artefact may be used multiple times, including as part of multiple architecture deliverables. Examples include an interface catalogue, location map, cross-reference table, context diagram and data model.

Entity: a discrete element that is part of the architecture of a system and is of interest to anyone performing architectural analysis and design. Entities may appear in one or more artefacts. Entities can also be related to each other by **abstraction**. For example, a retailer may have a large number of products of various types, such as tablets, computers and smart speakers. These may be generalised into a category called 'devices'. This category is an architecture entity as well, and is related to all the entities it categorises. Entities may be created by **composition**; several separate products may be sold together and become a new entity that is related to the other products that compose it. With both **generalisation** and composition, there is a relationship between entities that can be used for architecture analysis, such as in a view. Entities may also be related by being modelled together in an artefact. Examples of entities include stakeholder, business unit, location, activity and event.

ADs may be created and stored in multiple ways, for example by using separate documents with a shared storage location; however, this can be hard to manage. A more effective way is to use a repository that is specifically designed for the purpose. Repositories provide facilities to store the AD components (architecture deliverables, artefacts and entities) and the associations between them. Repositories may also provide version control and configuration management to handle the changes that will be made to ADs and their components through the process of solution architecture.

5.1.2 Version control and configuration management

Architecture, and in particular solution architecture, is an activity that may involve a large number of documents and artefacts, which are acted upon and modified by multiple people, both sequentially and in parallel. For this to work efficiently and effectively, it is imperative that people base their actions, decisions and designs on the **latest** and **most relevant version** of each document or artefact. This means that version control and configuration management are a critical part of any architectural process.

Version control and configuration management have an additional role in gap analysis that makes them an **intrinsic part of the method**. The two systems in combination can be used to determine which components and interfaces have changed between two ADs.

- **Unchanged components:** have the same ID and version number and are part of the same configuration units in both ADs.
- **Changed components:** have the same ID in both ADs, but the version number will be higher in the target or transition AD.
- **Components with changed interfaces:** have the same ID and version number in both ADs, but are part of a new or different configuration unit, or one with a higher version. Note that this indicates that one or more interfaces may have changed, but a change to a configuration unit does not necessarily affect the interfaces of all the components. This can be established by examining the details more closely.

5.2 GAP ANALYSIS IN THE SOLUTION ARCHITECTURE LIFE CYCLE

There are many occasions in designing the architecture of a solution when it is useful to make an assessment of the type or scale of change. This can be, and often is, done informally, based on the knowledge and previous experience of the solution architecture team or other stakeholders. Gap analysis provides a more formal method and is flexible enough to be performed at different levels of detail, depending on the situation and how precise and accurate the results need to be.

5.2.1 Business case options

A business case may contain a variety of information about the problem area and the solution being proposed. In the solution architecture life cycle, a major function of the business case is to present options for solving the problem and allow stakeholders to decide between them. Each option in the business case has a number of assessments associated with it, including:

- **Cost benefit analysis:** an estimate of the net benefit to the business.
- **Return on investment forecast:** the timescale over which the benefit will be achieved.
- **Risk assessment:** what is at stake in adopting or failing to adopt this option.
- **Gap analysis:** an estimate of the amount of change required to deliver this option.

The output from the gap analysis performed for each option in a business case provides useful information that can help to decide which of the options will be best for the business.

The level of detail of a gap analysis performed for a business case option is necessarily limited by a number of factors, including:

- Time available for preparation.
- Level of detail available about solution components and their interfaces.

- Amount of information that can be taken into account by stakeholders where multiple options are under consideration.

- The comparative nature of the information required to distinguish between options.

At this point in the life cycle, the requirement is to have sufficient information for stakeholders to make an informed choice between the available options. Sometimes a new hybrid option will be proposed that combines features and benefits from two or more of the options being considered. If this is the case, a new gap analysis (and other assessments) must be performed to allow stakeholders to make an informed decision about the new option.

Fallowdale Hospital patient communications: business case options

The following options have been included in the patient communications business case for consideration:

- **Option 1:** outsource email communication to a third-party cloud provider with a SaaS interface. Outsource text messaging to a government service (called Notify) with a REpresentational State Transfer (REST) API.

- **Option 2:** use a CRM system (possibly cloud-based) to manage all patient communication.

- **Option 3:** integrate email and other communication channels with existing postal delivery system.

Gap analysis reports are required for each of the options. Figure 5.3 shows the baseline architecture.

Figure 5.4 shows the gap analysis between baseline and option 1 (outsourcing).

Figure 5.3 Patient communication baseline architecture (high level)

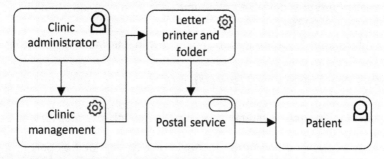

Figure 5.4 Patient communication option 1 architecture (high level)

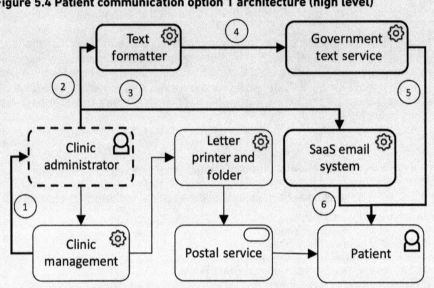

Gap analysis of components and interfaces between baseline architecture and option 1 architecture:

- **Clinic administrator (person):** modified role, increased responsibilities affecting capacity, training required, new interface (2) with text formatter, new interface (3) with SaaS email system.

- **Clinic management (system):** new interface (1) providing output for text and email.

- **Letter printer and folder (system):** unchanged with capacity reduced in future.

- **Postal service (service):** unchanged with capacity reduced in future.

- **Patient (person):** possible new communication methods (email and text), communication required.

- **Text formatter (system):** new component with interface (4) with government text service.

- **Government text service (system):** new component with interface (5) with patient.

- **SaaS email system (system):** new component with interface (6) with patient.

The high-level baseline and target models are also provided as part of the gap analysis report. The target model has been marked up with new components having thicker lines, new interfaces having thicker lines and numbers, and changed components having thicker dashed lines.

Activity 5.1

Perform a gap analysis for option 2 of the Fallowdale Hospital patient communications business case. The target architecture is provided in Figure 5.5. Note that the interface between the clinic management and letter printer and folder components has been deleted, as shown by the ⊗ symbol.

Gap analysis is used to compare business case options so the one that is best for the business can be selected by stakeholders. Apart from the list of changes to components and interfaces you have created, what additional information would help the stakeholders to select the best option for the business?

Figure 5.5 Patient communication option 2 architecture (high level)

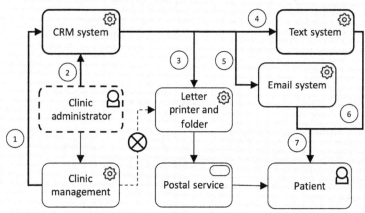

The gap analysis report for each option in a business case identifies the proposed changes to the high-level components and their interfaces, and presents them as a list. The number of items in the list gives an indication of the amount of change that would be involved in moving from the current state to the option under consideration. However, this can only give a very general indication of the overall size of the change. Some of the individual changes may be more difficult, expensive or disruptive than others, and this may significantly affect which of the options is best for the business.

One approach to improve the decision-making process is to extend each item in the list of changes to include some more detail about the nature of the change:

- **Work:** the resources and effort to bring about the change.
- **Cost:** the investment in components and interfaces and other known or estimated financial values.
- **Time:** how long each change will take to put in place.

Note that these details can be used to refine the cost benefit analysis, return on investment forecast and risk assessments for the proposed option.

Other details can be recorded, but the focus must be on providing information that is relevant to the selection of the best option for the business. More detailed information will be included in subsequent gap analysis reports that are to be prepared once a solution option has been selected and agreed by all stakeholders.

5.2.2 Logical design phase

Gap analysis is a useful tool that can be used during the logical design phase to assist in deciding between design options. As the options under consideration during this phase are only one part of an architecture, the gap analysis being performed here is on a smaller scale and has a narrower scope than the more comprehensive analysis of either business case selection or the validation phase.

However, having a narrower focus does not mean that the gap analysis is less detailed. Sufficient detail must be examined in order to make the correct decision between, for example, two components that could fulfil a particular requirement as part of the solution design. This can often mean drilling down to a lower level of detail to look at subcomponents and configuration settings, for example, to make the comparison sufficiently informative and to enable the correct decision to be made.

Fallowdale Hospital patient communications: selecting design options

At present, letters are printed by sending a batch instruction to the printer (see Figure 5.6). The instruction contains a job number that tells the printer what to do and a data set of names, addresses and appointment details.

Figure 5.6 Batch printing letters

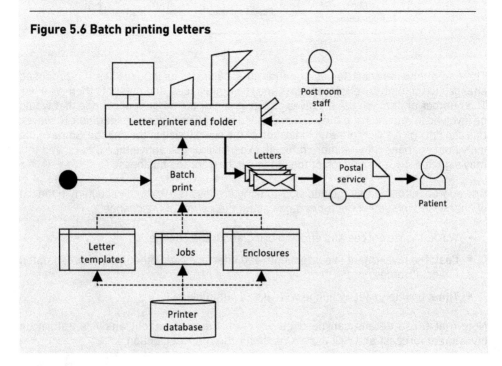

The solution architecture team is considering the best way to construct emails to contain the same information. The patient can then receive the invitation by either method of communication (or possibly both) based on their preference.

The printer software turns the data into individual letters based on templates and enclosures that are linked to the job number.

Option 1: copy the letter templates and enclosures to a new location and use a new mail merge component to construct emails.

Option 2: design a more flexible data structure to store message content and use a new component to construct emails.

A superficial gap analysis of these two options suggests:

Option 1: copy existing data to new data store and use mail merge component. Data storage and mail merge are currently in use within the hospital, although they would need to be configured for the new purpose and the capacity checked and extended if needed. The data needs to be copied but not modified. Overall assessment, small gap.

Option 2: design new data structures, extract, transform and load data. Create new message builder component. Overall assessment, medium gap.

What the gap analysis does not show is that option 2 is more flexible and will be able to handle more types of communication. Which option is finally chosen will depend on a number of factors such as appetite for change and urgency of delivery. Since these are alternatives, it may be that option 1 is put in place initially while option 2 is built for a later deployment.

5.2.3 Validation phase

Gap analysis is the last activity of the validation phase of the solution architecture life cycle. The validation phase follows the logical design phase and precedes the roadmap development phase. Gap analysis is a separate activity in the validation phase, whereas it is a supporting method for business case option selection in the analysis phase and in the logical design phase, where it is mainly used to support decision making.

Gap analysis plays a much bigger and more important role in the validation phase, however. There are four main reasons for this:

- **Scope of architecture:** the changes that are in scope for solution architecture include all changes that are required for the solution to deliver the agreed set of requirements, and the benefits from the business case. The scope is defined during the analysis phase immediately after the solution option has been selected.

- **Level of detail:** much more detail about the components and interfaces that make up the solution architecture is available for analysis at this point, because the activities of building block analysis and interface analysis have been completed and their respective catalogues produced.

- **Maturity of design:** logical design takes the outline solution and all the other inputs, requirements, constraints, building block catalogue and interface catalogue, and constructs the best possible design that maximises benefits, minimises impact and is open to future changes as far as possible. After the logical design phase, the design is validated through the use of views and viewpoints and is tested against enterprise architecture models. These activities serve to harden the design towards a fully mature state. The impact analysis, which is performed immediately before gap analysis in the validation phase, may also lead to minor design refinements so that the design is as mature as possible by the time the gap analysis activity begins.

- **Importance of output:** the list of individual changes to components and interfaces, and any dependencies between them that have been identified, are the basis for the delivery of the solution by P3M and solution development teams working to achieve the delivery roadmap.

The outputs of this gap analysis are more formal and subject to review by the solution architecture team, P3M and other stakeholders. The outputs are also much more comprehensive and detailed because of the scope and depth of the analysis and will include:

- **Gap report:** larger, containing more components and interfaces with more detailed information about the changes required, and exposing a lower level of detail, where relevant, such as subcomponents and configuration options.

- **Gap models:** more extensive, including multiple levels that are required to perform the analysis in sufficient detail.

- **Views:** included because they show the changes from a variety of stakeholder perspectives and address specific concerns.

- **Enterprise architecture models:** included because they show the changes from the perspective of enterprise architecture and domain architectures such as business, data and applications.

5.3 INPUTS TO GAP ANALYSIS

In simple terms, the minimum inputs to gap analysis are two ADs that represent the before and after states. The two ADs can be detailed or high-level and have a broad or narrow scope.

For example, an AD can describe an entire enterprise architecture but at a high level with low detail. In this case, the gap report will reflect the inputs and contain high-level components and their interfaces which span the enterprise. If gap analysis is performed at this level, it is likely to be the starting point for a series of far-reaching business change programmes. Each of these high-level changes will require more detailed analysis so that the specific actions that will achieve the desired change can be identified.

5.3.1 Solution architecture models

The main expression of any type of architecture is through the use of models. In solution architecture, models are used throughout the design life cycle, some for investigating the problem or deciding between design options and some for capturing design decisions that will become part of the logical design architecture.

Gap analysis is used at these points in the solution architecture life cycle:

Business case option selection: the target architecture is high-level, similar to the solution outline or conceptual architecture produced immediately after the discovery phase of the solution architecture life cycle. Usually a similarly high-level model will be produced to represent the baseline architecture so that the business case options can be compared with it. This will usually have been done early in the life cycle, perhaps even begun during the initiation phase to aid understanding of the problem area. Such models usually evolve as more information comes to light, which, of course, is the main purpose of the discovery phase.

Logical design: here it is used to distinguish between design options and normally focuses on a small area of the architecture. The models used are therefore more limited in scope. These models usually encompass a single component and any components directly connected to it, including the interfaces that connect them. These small-scale ADs, which often comprise a single artefact, are constructed by the solution architecture team who are familiar with the components and interfaces involved.

Validation phase: the solution design that is produced during the logical design phase and tested and refined during the validation phase becomes the target architecture, or AD, for the solution. The mandatory gap analysis activity, which is performed at the end of the validation phase, uses this AD as an input; in fact it is the most critical input, as it represents everything that the business wishes to achieve through the implementation and deployment of the solution.

Gap analysis relies on the fact that the target AD can be compared with a baseline AD. If the organisation or enterprise has a well-developed, mature and fully integrated enterprise architecture practice, then a baseline AD for the current state of the solution being designed may be available. However, there are several situations in which such a baseline AD does not exist, including:

- **Legacy:** the current state predates the introduction of enterprise and solution architecture or has not yet been documented.
- **Greenfield:** the business area (or perhaps the entire business) is new and there is no existing solution in place.
- **Consolidation:** several existing solutions are being combined into a single one; ADs may exist for one or more of the existing solutions.

In fact, it is comparatively rare for a newly designed solution to be an exact match for a current one, and for there to be an existing AD available. This means that care must be taken when performing the gap analysis so that existing components and configurations are not omitted from the analysis. This could result in unnecessary activities being identified or opportunities for rationalisation being missed.

If there is an existing undocumented solution that has no AD, or several existing solutions that overlap with the target architecture, then a baseline AD must be constructed so that can be compared with the target AD with gap analysis. This is a fairly common situation. A good starting point for this is in the enterprise architecture repository or continuum. The models and other artefacts that are in scope for the solution are contained in the architecture inputs catalogue that was produced in the discovery phase. However, this may not be a complete list as the solution scope may have changed since that point, and a review is highly recommended.

In the case of a greenfield solution, there is no baseline and so every component or interface is new. However, some components may exist as part of other unrelated solutions and these can be found in the enterprise architecture repository or continuum.

5.3.2 Enterprise architecture models

Enterprise architecture artefacts that are within the scope of the solution or problem area are initially catalogued in the discovery phase. Many of these artefacts are models or include one or more models alongside descriptive text such as commentary and analysis. These artefacts and the models they contain, by definition, represent the current state or baseline architecture of the area they cover.

These models can be used to replace any missing baseline solution architecture models. They can even be used to construct the complete baseline solution architecture in the case of a completely new greenfield solution. This enables gap analysis, which would otherwise be impossible, because a baseline AD is a mandatory input for the method.

For example, information about data and information currently in use is contained within the enterprise data architecture. This can be filtered to match the scope of the solution so that it represents the baseline solution data AD. A similar process can be followed for business processes, applications and other baseline architecture artefacts.

Where a solution AD is produced from enterprise architecture artefacts, it may be done at various levels of detail during the solution architecture life cycle. For example, a high-level AD is required for the gap analysis that is used for business case option selection. A more detailed and comprehensive AD is required for the large-scale gap analysis that occurs at the end of the validation phase and immediately before roadmap development.

Another use of enterprise architecture models is to identify components that already exist but are not in scope for the baseline solution being replaced, redesigned or extended.

Fallowdale Hospital patient communications: existing components

The following components exist within the hospital's enterprise architecture but are not within the scope of the current patient communications solution:

- **Existing email capability:** only used for manual communication between employees within the hospital and ad hoc communication with external organisations such as GP surgeries and other healthcare providers. There is no automation, although the current system has some facilities for this.

- **Existing text communication system:** a text message system is being used for very specific communication to limited groups of patients. It is unclear whether this could be used for more widescale patient communication.

- **Other social media:** a presence has been established on various platforms; this may be of interest in the future.

- **PR unit:** skilled public communications staff exist within this business unit and this could be relevant for designing patient messaging.

Activity 5.2

Of the four existing components identified from enterprise architecture models in the Fallowdale Hospital patient communications case study, which are relevant to gap analysis, either for business case option selection or the main gap analysis activity during the validation phase?

Give a brief explanation for your answers.

5.3.3 Supporting documentation

Many types of documentation exist within an organisation or enterprise that may have a bearing on gap analysis, and it is not possible to give an exhaustive list. However, it is useful for the solution architecture team performing the gap analysis to be aware of documentation relating to components and interfaces that are included in the gap analysis.

This type of documentation can include:

- Descriptions of how components are currently being used, and any policies and procedures that constrain their future use.

- Technical documents giving specific details about components and their interfaces, such as configuration options.

- Reports, comments and analysis by the business that can assist in assessing the changes that are required to implement components and interfaces within the wider business context.

5.3.4 Viewpoints

Viewpoints would seem to be extremely useful for gap analysis, as their purpose is as a template for analysing architectures at different stages. For example, a viewpoint can be used to analyse the baseline and target architectures of a solution, resulting in two views showing the situation before and after the solution is deployed.

However, although views (produced using a viewpoint) can show the effect of change, they do not necessarily help to see the gap between two architectures.

Views can be helpful when deciding between business case or logical design options as they can highlight significant aspects of each choice. They can also provide additional information, especially about mitigation of loss when trying to balance the impact of change between the competing concerns of different stakeholder groups. Gap analysis can be performed with multiple alternative target architectures to assess which has the biggest change or impact, both of which will require management resources.

5.3.5 Additional information from the business

Sometimes it is necessary to obtain information from the business that is not contained in an existing AD or in any supporting information. This could be to obtain:

- clarification of requirements;
- details of costs, contracts and licences;
- solution development resource capacity;
- priority of the delivery of specific benefits.

These items of information may be used to improve accuracy when assessing changes either for decision making or when preparing the gap report that will feed into the delivery roadmap.

5.3.6 Compatibility and currency of inputs

Gap analysis is facilitated by using the same notation and format of the models being analysed. Comparing like with like helps to make the results more accurate, reliable and easier to interpret. For example, if the current state model is in the form of a table of facts and figures but the future state is represented as a map, then more work will be required to identify the gaps, possibly including converting models from one form to another. The extra work involved could introduce errors, which would make the results less accurate and reliable. If information needs to be categorised, for example when grouping geographical data from a map into tabular form, then this can make the results harder to interpret.

It is also clear that inputs must be up to date and only the currently authorised version of any input used. Models and other artefacts that make up the inputs to gap analysis commonly go through many revisions during their lifetime and during the solution architecture life cycle. Therefore version control is critical. For example, if a model is used that includes changes that were proposed but then withdrawn, for whatever reason, then the version of the solution design should be rolled back to the version without the withdrawn changes. Using the wrong version here would result in incorrect and unnecessary changes being included in the output of gap analysis.

5.4 STEPS IN THE GAP ANALYSIS METHOD

The simplest form of the gap analysis method compares two architectures and reports the gap between them. Each of the two input architectures can be as small as a component or as large as an enterprise. A solution AD is somewhere between these extremes.

The output can also vary, depending on what is needed. The output always has an assessment of the change in terms of size, but more detail can be added if it helps to make decisions and determine future activities. For example, gap analysis is used to decide between two solution subsystems in the logical design phase. Here, the size and complexity of change are the deciding factors. Any detail about the steps to achieve the change is not required until the validation phase. The output therefore needs to be simple and will only be documented, if at all, as a design note.

In the solution architecture life cycle, the complete gap analysis process with detailed output only needs to be followed at the end of the validation phase. The detailed outputs are needed to inform the building of the delivery roadmap.

5.4.1 Verify and validate inputs

Even if a simple gap analysis is being performed, it is important to use the correct inputs, otherwise the results cannot be relied upon. This can lead to the wrong decisions being made and incorrect information being communicated to team members and the business.

Inputs must first be confirmed to be the correct and relevant AD or architecture element. In addition, the version number, date and status must be checked to ensure the latest and currently-in-use inputs are being used. This can be more complicated than it sounds, as there may be many elements within an architecture, such as artefacts and entities. If any of these is not the current version, the results are unreliable and may be inaccurate.

The version and configuration details of the inputs being used should be recorded and appear in any reports or notes that form the outputs of gap analysis, which themselves need to be under version control. This is very important and can help to reduce the risk of using the incorrect version in future.

5.4.2 Rationalise for consistency and completeness

In order for two architectures to be compared using gap analysis, the inputs used need to be consistent and compatible with each other.

Inconsistency of inputs can take the form of:

- scope, such as excluding part of the solution or including irrelevant elements;
- type or format of documents;
- level of detail.

Inputs may also be incomplete if any elements of the architecture are missing.

To correct for any of these factors and allow the gap analysis to go ahead, the input architectures can be cross-referenced with available solution architecture artefacts such as:

- solution scope;
- component catalogue;
- interface catalogue;
- logical design.

5.4.3 Grid method

A grid method for architectural gap analysis has been proposed by various authors (Tannady et al., 2020; TOGAF 9.2, 2018) that compares the components of two architectures (see Figure 5.7). This method places the components of one architecture (usually the baseline) on the vertical axis and those of the comparison architecture (usually the target) on the horizontal axis. Changes to components in both architectures are annotated in the intersection cell. A row is added so that newly added components can be annotated, and a column is added for those that are to be deleted.

This is a visual method and the annotations being in grid cells has the advantage of clarity as to which components they refer to. It is also easy to see cells that should contain information but are empty:

- **Unchanged:** intersection cell indicates no change.
- **Modified:** intersection cell contains details of change.
- **Added:** cell in bottom row below component contains details of addition.
- **Deleted:** cell in right-hand column in line with component contains details of deletion.

The disadvantage of this grid arrangement is that many of the grid cells remain empty and unused. Only the diagonal intersection cells are used for details of changes, and the bottom row and final column are unused except for new and deleted components respectively. Unlike other methods of analysis that use a grid, for example cross-referencing between data and applications, in gap analysis an empty column, row or cell in a grid is not a helpful visual indicator that there might be something missing. If it is not an omission, then it serves as a useful confirmation that no change is required.

Figure 5.7 Gap analysis grid

		Target architecture			
		Component 1	Component 2	Component 4	Deleted
Baseline architecture	Component 1	Details of any changes			
	Component 2		Details of any changes		
	Component 3				Details of deletion
	Added			Details of addition	

5.4.4 Table method

The table method is a relatively simple approach that is more suitable for small-scale solutions with few components that need to change. It involves making a list of all components and interfaces that are part of both input architectures (see Table 5.1). The list forms the first column in the table and must be de-duplicated so that each item is unique.

Subsequent columns may be varied to suit the purposes of the gap analysis, but the following are commonly used:

- **Name:** name of component or interface.
- **Change type:** whether changed, unchanged, added or deleted in the target architecture.
- **Details of change:** relevant information for the user(s) of the analysis.
- **Internal impact:** cross-reference to any other items in the table that are affected by the change described.

Table 5.1 Gap analysis using table method (extract)

Name	Change type	Details	Internal impact	Costs	Resources required	Timescales	Risk factors
Patient Administration System	U						
Clinic Administration System	M, I	New FR: • Rq037 • Rq172 • ... Remove: • Tk003 • ... New interface: Message System: • Rq096 • ...	Staff training: Clinic Administrator Volume: Patient Administration System Network capacity ...	4,300	Supplier consultancy Internal 47 person-days	30 days elapsed	Access to patient data (Rk209) Reliance on automation (New) Data currency and accuracy (Rk573)
Clinic Administrator	U						
...	...	...	...	...	...	...	...

U=unchanged, M=Modified, I=New interface, N=New component

FR=Functional requirement, Rqxxx=Requirement reference, Tkxxx=Task reference

Rkxxx=Risk management reference

146

Other columns may include:

- Cost implications.
- Resource requirements.
- Timescales.
- Risk factors.

This method has the advantage of simplicity as well as the benefit that the table used for analysis also contains the majority of the output required for a gap report. Any models that are used in the analysis can be provided in the output to support the information in Table 5.1.

5.4.5 Graph method

This is a visual approach to analysing the differences between architectures. It is based on two compatible models that may exist prior to the analysis or be built specifically for the purpose. Each of the two input models is a graph where the nodes are components and the lines that connect them are interfaces. Gap analysis compares each element of the two inputs to produce a third graph that becomes the output. All elements are represented in a way that makes it possible to determine the scale of change.

The graph method has the advantage that the output model highlights interdependencies between elements and makes visible how a change in one element impacts on other parts of the architecture.

Activity 5.3

Perform a **graph method** gap analysis based on the baseline and target graphs shown in Figures 5.8 and 5.9.

Produce a graph that highlights the changes.

Show new elements with thick lines, modified ones with thick dashed lines, and components with new interfaces with a lollipop symbol (–o).

Figure 5.8 Baseline architecture graph

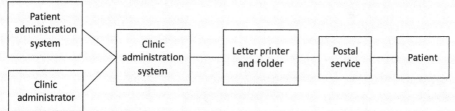

Figure 5.9 Target architecture graph

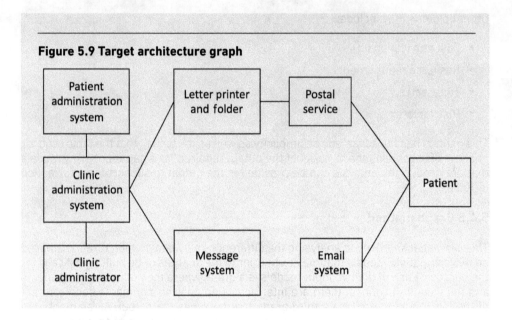

5.4.6 Automated gap analysis

It is possible to fully or partially automate the activities of gap analysis. This relies on the details of both input architectures being available in machine-readable format, such as a structured architecture repository or other data store. The graph method, for example, can be automated if architecture descriptions are stored, or can be easily converted to be stored in a graph database.

Checking for compatibility and completeness relies on effective version control and configuration management because this is how the gap analysis algorithm can identify changes.

In theory, this could be a useful approach. However, it is impractical in many situations, either because of the immaturity of enterprise repositories or of the architecture practice of the organisation or enterprise. If too much manual work has to be done to support the automated process, the benefits are hard to sustain.

5.4.7 Itemise components and interfaces

Whichever method is used, the output must itemise the elements of both architectures and provide a commentary on each item. Depending on the analysis method used the list may need to be:

- extracted from various gap models;
- rationalised to ensure naming consistency;
- de-duplicated so information is collated.

Note that if the table method has been used then there is little or no work to do as all the information is available in list form.

Once the list has been rationalised, the information obtained by analysis can be recorded in the correct place, next to the component or interface to which it relates.

5.5 OUTPUTS FROM GAP ANALYSIS

There can be multiple outputs from a gap analysis depending on its type, purpose, and when it occurs in the solution architecture life cycle. The simplest output is a written opinion about the size of the gap, which might accompany a design decision note during logical design, for example.

If a formal technique has been used or models prepared as part of the gap analysis, then these are often useful as outputs as they can expand on how the conclusions were reached and can also be valuable for future gap analyses. For example, where a design option was rejected for inclusion in an intermediate architecture, it may be relevant when that option is included in a target architecture at a future stage. All the outputs of gap analysis are lodged in the enterprise architecture repository with the other artefacts of solution architecture. One of the main uses of outputs in the solution architecture life cycle is in the creation of the delivery roadmap – the basis for planning the delivery and deployment of the solution.

5.5.1 Gap report

One essential output of gap analysis is a list of architecture **components** and **interfaces** that differ between the two models being compared. Components and interfaces are both architecture **elements**.

Each element that will change is itemised and identified as either an addition (new element), deletion (component no longer required) or modification (component will still exist but in changed form).

This itemised architecture element gap report acts as a catalogue of changes that need to be made to transition from one architecture to another.

When the time comes to implement the solution, these individual changes will appear in programme and project plans. Such plans take into account the dependencies that exist between activities and, if this is known about when writing the gap report, it should be recorded.

Apart from the list of changes, for a gap report to be useful it needs to include additional information. For example, to use the gap report to select an option proposed in a business case, it is necessary to estimate the general scale of change including cost, time and resources, and the details of these estimates should be added to the gap report. At a later stage, when plans are being drawn up to manage the change, there will need to be more detail about how the change to each architectural component will be achieved, including the steps and actions that will be taken. At this later stage, the items that were previously estimated can be refined and made more accurate and concrete.

There is a separate activity that looks at the impact of a solution on the broader architecture landscape. This activity is called **impact analysis** and produces a report on the zone of impact – that is, areas outside the scope of the solution that are affected by its deployment. There is, however, another category of impact in which an element of solution architecture affects another element. These **internal impacts** should be recorded in the gap report if they have been identified. This will also be useful when preparing project and programme plans.

Fallowdale Hospital patient communications: internal impacts

The following internal impacts have been identified within the existing solution space:

- **Post room staff:** fewer letters will be sent with the new solution so the staffing requirements should be reduced; this impact needs to be assessed.

- **Patient administration system:** new access controls must be implemented to meet the security policies about personal information.

- **Network:** encryption settings must be updated to comply with security standards.

- **Business processes:** several business processes must be updated and new service levels agreed.

5.5.2 Gap models

A very useful artefact that can be produced during gap analysis is a gap model. This is essentially a combination of the baseline and target versions of a model, artefact or viewpoint (see Figure 5.10). The idea is to show the transition using a model and visualisation that stakeholders are familiar with, or at least will recognise. All architecture elements are shown on the model, including those that will remain unchanged as well as additions, deletions and modifications (see Figure 5.11). Commonly, these are colour-coded with additions highlighted in, say, green, deletions in red, and so on. Gap models can be highly valuable stakeholder communication tools as they are able to focus on a specific area of interest and benefit from their resemblance to existing familiar models that were designed to present information clearly and concisely. They are essentially a combination of the 'before' and 'after' representations designed to show the transition between the two.

In addition to one or more gap models that are produced to help communicate the change to stakeholders, any models that were developed specifically as part of the gap analysis method – for example, a new baseline model of a legacy solution – are legitimate and useful outputs and should be provided as such.

Figure 5.10 Baseline and target models

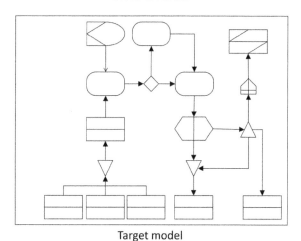

Baseline model

Target model

5.6 ENTERPRISE ARCHITECTURE ARTEFACTS

A number of enterprise architecture artefacts may have been used as inputs to the gap analysis method. For example, the enterprise data architecture may have been used as a reference point for designing new data and information elements of the solution. If the solution design changes anything about the enterprise data model – for example, adding new data structures that are needed by the solution – then the enterprise data architecture will need to be updated.

Gap analysis is performed before the implementation of the solution, so there is an argument for delaying any changes to the enterprise data architecture until the point

Figure 5.11 Gap model showing changes

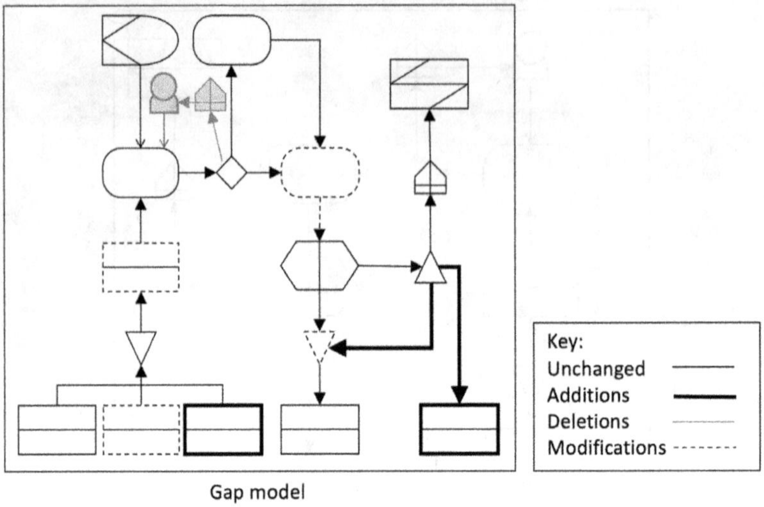

Gap model

at which the solution is to be deployed. However, solutions do not exist in a vacuum, so enterprise and data architecture functions need to be aware of the changes, as they may have an impact elsewhere – for example, another solution may be being designed or built whose scope overlaps with the enterprise data architecture elements being modified.

Any enterprise architecture artefacts that are modified by the solution architecture process become outputs for gap analysis. This is because updating the repository is a change activity that needs to be completed by enterprise architecture as a result of the solution design.

5.7 PREPARING FOR THE ROADMAP DEVELOPMENT PHASE

Simply having a list of what needs to change does not give any guidance as to how to bring about the change. However, the advantage of the architectural approach is that the changes are broken down to component level and categorised, which is a good starting point. A typical next step would be to assess the amount of time, specialist capability, finance and other resources needed to achieve the change to each item in the list. Additional factors such as risk and interdependency may be recorded.

The delivery roadmap shows when specific benefits will be delivered to the business. Benefits are linked to the implementation and deployment of architecture elements that are itemised in the gap report. The programme and project plans that are used to manage the delivery of the architecture elements and the benefits they enable rely on accurate and complete information about what is needed to deliver them and what dependencies exist between them.

The output from the detailed gap analysis performed in the validation phase therefore provides an invaluable structure and an initial estimate of the information required for planning. More accurate figures and other details of resources and time will be needed when developing programme and project plans, but in many cases the gap report will be sufficiently detailed to develop the delivery roadmap.

REVIEW QUESTIONS

1. When can the gap analysis method be used in the solution architecture life cycle?

   i. When identifying stakeholder concerns.

   ii. During the logical design phase.

   iii. To help build and validate viewpoints.

   iv. As part of business case option selection.

   v. Immediately before the roadmap development phase.

   a. ii, iv and v only.
   b. i, ii and iii only.
   c. i, iii and iv only.
   d. iii, iv and v only.

2. What input to gap analysis provides information about the current state?
   a. Consolidation architecture description (AD).
   b. Target architecture description (AD).
   c. Baseline architecture description (AD).
   d. Transition architecture description (AD).

3. Which output of gap analysis itemises the changes required to implement the solution that has been designed?
   a. Change catalogue.
   b. Gap models.
   c. Roadmap.
   d. Gap report.

4. A solution architecture team has performed a gap analysis, but a business stakeholder notices that an important solution component is not included in the output. What is a possible explanation for this?
   a. Stakeholder concerns were incorrectly recorded.
   b. Version control was not applied correctly during gap analysis.
   c. The delivery roadmap contains errors.
   d. The solution component is to be added during the completion phase.

6 STAKEHOLDER INTERACTION

LEARNING OUTCOMES

When you have completed this chapter, you should be able to demonstrate an understanding of the following:

- The stakeholder categories involved with solution architecture
- Stakeholder identification and evaluation
- Stakeholder roles in key activities of the solution architecture life cycle
- Managing concerns and defining viewpoints
- Business case option selection
- Scope definition
- Validation and sign-off of a chosen solution
- Governance and design authorities

6.1 SOLUTION STAKEHOLDER CATEGORIES

Solution stakeholders are the connection between a solution and the business. If correctly selected, they represent the business and can make decisions on its behalf. If the solution satisfies the stakeholders, it will be what the business wants. This is underpinned by the fact that all the decisions made during its design are made by and owned by the business.

The word stakeholder has a very broad definition: 'A person with an interest or concern in something, especially a business' (Oxford Dictionary, 2021). More specifically, ISO 42010:2011 defines stakeholder as an 'individual, team, organisation, or classes thereof, having an interest in a system'.

(ISO/IEC/IEEE 42010:2011, 2011)

A solution stakeholder is any person or group with an interest in, or concern with, the solution.

The solution architecture team is responsible for ensuring that the correct stakeholders have been identified and for facilitating appropriate communication with and between stakeholders.

Stakeholder communication is required for the following reasons:

- **Information:** ensures everyone is aware of developments, mainly one-way but with the opportunity for feedback.
- **Consultation:** captures business expertise relating to the solution; this is two-way and proper records are essential for traceability of input and decision making.
- **Accountability:** records the approval of architecture decisions and actions by specific stakeholders.
- **Responsibility:** allocates and tracks tasks that are given to stakeholders to manage and complete, possibly with the support of P3M.

Communication often involves the solution architecture team, but there may be interaction directly between stakeholders. Record keeping is critical wherever and however design decisions are being made and is the basis on which all stakeholders can be kept fully informed.

From this analysis, solution stakeholders are people or groups who:

- need information about the solution;
- provide essential business or domain expertise;
- have authority for budget, resources or other organisational decisions;
- take part in the design, implementation or deployment of the solution.

This can be used as a checklist to ensure all stakeholders are included.

6.1.1 Business owners and senior managers

Business owners and senior managers have the ultimate responsibility for the success of the business. They are able to make strategic decisions that are in the best interests of the business. In large organisations or enterprises they may not have sufficient knowledge of the detail of day-to-day operations to make all decisions, and these may be delegated to those who are closer to the operation of the business.

Businesses may be owned by:

- Individuals in private companies and partnerships.
- Shareholders in public listed companies.
- Another organisation, for example in the case of a subsidiary or investment company.

A business may be a commercial organisation trying to make a profit from its activities, but there are also not-for-profit organisations such as those in central and local government, health and social care, and the charity sector.

Some owners have executive decision-making powers within an organisation, but this is usually given to a broader group such as a board of directors that includes the senior managers of the organisation.

The concerns of business owners and senior managers regarding solutions include:

- Meeting strategic business goals.
- Improving the competitive position of the business.
- Improving the business' financial position.
- Improving the business' market position, in terms of being the market leader or leading-edge innovator.
- Strengthening the organisation's resilience to external change.

Fallowdale Hospital patient communications: business owners and senior managers

The hospital is publicly owned but operates as a stand-alone entity with a board of directors:

- Chief executive.
- Chair.
- Director of strategy.
- Finance director, in charge of IT.
- Operations director, in charge of administrative staff in all areas including Outpatients.
- Medical director, in charge of all doctors and medical specialties.
- Director of nursing, in charge of inpatient, outpatient and specialist nurses.
- Three non-executive directors who represent the community served by the hospital.

The hospital is accountable to a government department that sets the national healthcare strategy and receives operational reports. The department has an input to strategy and provides guidelines for care provision but is not involved in day-to-day operations.

6.1.2 Business sponsor or product owner

Within an organisation, a single person is usually chosen to be the public face of a solution and given the authority to ensure its success. This role has been called **business sponsor** by P3M and overlaps with the role of **product owner**, which is part of the **Scrum Agile framework**. It is worth noting that the term **service owner** is also used for this role, and may be more appropriate for solution architecture, as a solution often involves delivering a holistic service rather than a single product.

The Scrum Guide (Schwaber and Sutherland, 2020) says that the 'Product Owner is accountable for maximizing the value of the product resulting from the work of the Scrum Team', and adds that they must be respected as the decision-making authority (for the solution) throughout the organisation.

The Association of Project Managers (Murray-Webster and Dalcher, 2019) describes a successful project sponsor as 'A leader and decisionmaker who is able to work across corporate and functional boundaries within the organisation'.

This person is therefore the most important stakeholder for solution architecture and must approve or delegate all decisions about the solution.

The concerns of the business sponsor or product owner are that:

- All the solution requirements will be met.
- The problem can be solved and the proposed solution will succeed in doing so.
- Risks and reward are balanced in the solution design.
- The chosen solution is achieved with the optimum budget of time, money and use of resources.
- Any line of business benefits are the right ones with the right priority.
- Communication to the organisation is effective and accurate.
- Nothing is done that they would disagree with.
- Their personal time and energy are used efficiently and the whole process is not too time-consuming.

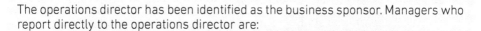

Fallowdale Hospital patient communications: business sponsor

The operations director has been identified as the business sponsor. Managers who report directly to the operations director are:

- head of outpatients administration, including clinic management;
- print room manager;
- procurement manager.

The majority of the end users of the patient administration system work in departments that are the responsibility of the operations director.

6.1.3 End users and business actors

The term **end user** is usually applied to people who directly interface with **software**. People who play a part in the **delivery** or **enablement** of **business services** are called **business actors**. End users of any software that is part of a solution are therefore also business actors for the solution and any business services it delivers or enables.

A business actor may be at any level of seniority within the organisation or enterprise but must play an active role within the solution.

End users and business actors have the greatest involvement with the solution. In many cases, this is the stakeholder group most affected by changes to a solution.

They can contribute to the solution design by providing information about the current situation and advising on how any proposed solution would work. They often have specialist information about the line of business or knowledge domain in which the solution operates.

The concerns of end users and business actors are:

- The new solution improves the service they can deliver, and they are trained in how best to deliver it.
- Everyone understands their roles and responsibilities and any major changes requiring HR negotiations are anticipated and dealt with fairly.
- Implementation does not cause unnecessary disruption to normal operations and the change is well managed.
- Feedback from those who do the job is taken seriously throughout design and implementation.

Fallowdale Hospital patient communications: end users and business actors

The end users and business actors for the patient communications solution are:

- Medical department staff, who organise and run clinics, manage referrals and put patients on waiting lists.
- Outpatients administrators, who manage the process of inviting patients, recording attendance at clinics and booking follow-up appointments.
- Post room staff, who manage the letter printing process once a batch request has been received through to handing the finished letters over to the postal service.
- The public relations (PR) department currently manages large-scale communication programmes and has expressed an interest in using the patient communication system. The PR department has responsibility for hospital-wide messaging. This is driven by a combination of hospital strategy, urgent medical alerts and other local and national initiatives that require

communication with patients, other service users and other healthcare provider organisations. The PR department may be given responsibility for all patient communication, including the content of individual patient communications.

- Patients are currently customers or service users. They may become end users if a system is set up for them to manage their communications preferences.

6.1.4 Customers and business service users

Customers of the business are the focus of business activity and those for whom all business services are designed. Many solutions aim to improve some aspect of the quality of the service given to customers, such as offering more choice, faster turnaround or better information. Some solutions are driven by customer feedback or competition for customers. Where customer behaviour is critical to the success of a solution, it is vital that they are engaged with, and have a high degree of interaction during, solution design.

Some solutions are for internal use and do not have any direct interaction with customers of the business. The internal users of a solution are sometimes described as internal customers or service users, although the distinction between business actor and internal customer has more to do with organisational structure than solution architecture. Internal solutions may have little direct or visible impact on external customers. For example, a solution for improved staff allocation or to reduce the cost of processing is mainly of concern to internal business stakeholders.

The concerns of customers and business service users are many and varied, but some that are frequently found are:

- Contracts and agreements must be met.
- Customer service should be consistent and exemplary.
- The service provider should give a quick response to any request.
- The business must be able to offer and provide the service needed.
- The interaction must be coordinated so that everyone involved in delivering the service is fully aware of the situation. This is sometimes referred to as 'joined-up thinking' or 'left hand knowing what the right hand is doing'. Customers do not want to be asked for the same information multiple times.
- The service must be provided in a way that is convenient for the client, for example 24/7 contact with online or messaging if appropriate.
- Communication during design and implementation must be clear and appropriate, with updates or alerts when changes occur or action is needed.

Note that these largely relate to the business service that is the main point of interaction between the customer and the organisation or enterprise. The business service may not be identical in scope to the solution, but acts as a proxy for it as far as the customer is concerned. As with all stakeholder categories, customers may have very specific concerns that depend on the type and nature of the solution. The concerns shown here are general ones that are common to many solutions.

Fallowdale Hospital patient communications: customers and service users

The main customer or service user of the patient communications solution is the patient. The service they use is consultation with, and sometimes treatment by, a healthcare professional in an outpatients clinic. The patient communication system is the mechanism for communicating invitations and supplementary information from the hospital to the patient. Communication can happen in the reverse direction but only if the patient wishes to change or cancel the appointment. This reverse communication is currently by phone during limited opening hours.

The initial design of the patient communication solution will be limited to outpatient invitations. This may be extended to include other types of communication with the patient in subsequent designs. The baseline uses letters sent via the postal service. The initial solution design adds email and is considering text as options for communication. This may also be extended to include other means of communication. The patient would still be the customer or service user with these extensions.

The patient communication solution may be extended to include external healthcare professionals. If so, these healthcare professionals would become customers or service users of the communication solution. The solution would need to be renamed if this extension comes into scope.

6.1.5 Solution architecture as a stakeholder

It seems strange, or perhaps too obvious, to include the solution architecture function as a stakeholder. However, given its responsibilities for designing the solution and role in governing the implementation and deployment, this is hardly surprising.

A solution architecture function or team may include:

- solution architects;
- business analysts;
- systems designers;
- data and information modellers;
- other specialist roles.

A common working model is to assign a solution architect during the initiation phase, who will request additional resources that will vary throughout the life cycle.

Solution architecture has the following concerns:

- **The solution is successful:** it is able to solve or at least address the problem, which is compatible with the business's strategy, ethos, principles and appetite for risk.

- **Efficient use of resources:** a small change that can produce big benefits is preferred to a large-scale programme of activities with little impact.
- **Reputation of solution architecture:** the benefits to the business are clearly visible, meaning that solution architecture is more likely to be consulted and used when problems are encountered in the future.

6.1.6 Enterprise architecture and subdomains

The enterprise architecture function has a high-level, broad perspective of solutions and how they fit into the enterprise as a whole. New solution designs must fit within the existing architectural landscape and any conflict or duplication will cause problems for enterprise architecture, either immediately or when future change is planned.

Enterprise architecture is usually broken down into subdomains:

- business architecture;
- applications architecture;
- data or information architecture;
- infrastructure or technical architecture;
- security architecture.

These subdomains have the same scope as enterprise architecture and new solution designs must be compatible with each domain separately, as well as with enterprise architecture overall.

Enterprise and subdomain architectures are involved in solution design, mainly in a consultative capacity. For example, data or information architecture can advise on the design of data and information structures in a solution; applications architecture can advise on whether any existing applications can provide functionality needed to satisfy a solution requirement. All architecture functions have a governance role, for example during business case option selection, and beyond design, during implementation and deployment.

Concerns for enterprise architecture and subdomain architecture functions include:

- Upholding and enhancing directives.
- Improving consistency of design.
- Taking advantage of opportunities for consolidation.
- That solutions should not conflict with other change initiatives without the opportunity for mitigation.
- That no work should be duplicated between solution and enterprise architecture.
- That the solution does not cause unexpected side effects that have not been accommodated or mitigated for.

Enterprise and subdomain architecture stakeholders are involved in the design. Adequate processes need to be in place so that their concerns about solution architecture and the designs produced are accommodated.

6.1.7 Stakeholders involved in solution design

Apart from enterprise architecture and its subdomains, other stakeholders are involved in solution design. Several roles are listed here but they do not necessarily represent separate people, as roles are sometimes combined or may not be present at all. These include:

- **Software architecture**: advises on software capabilities, interfaces and integration with other components in solution design.
- **Business analysis:** represents the business with detailed and systematic knowledge of specific areas such as lines of business or strategic aims.
- **SMEs:** provide details of the content and behaviour of specialist domains and areas of the business, and external factors that impact on solution design, such as customer behaviour.
- **Technical specialists:** provide details of standards and technology, such as in the area of security, where these are specified in the solution requirements.

Their concerns include:

- Their capacity should be used effectively and not wasted.
- Specialist focus must be represented to stakeholders for them to make informed decisions.
- Specifications are suitable for designing and building components and interfaces.
- Specifications correctly represent the specialist interests of the contributors and those they represent.

Note that some of these specialist roles may be involved in implementing the solution by, for example, designing software, re-engineering business processes and procuring technology. If so, they will share the concerns of others who are involved in solution implementation and deployment.

Fallowdale Hospital patient communications: stakeholders for solution design

The following groups will potentially be involved in the design of the patient communication solution:

- **Software architecture:** clinic management software will be modified to send invitations (and other messages in future) based on patient preferences. If messages are being sent by email or text then software components and interfaces will be required. Depending on what design decisions are made about storage and use of the message contents for alternative communication media, some new software components and interfaces will be required here. Software architecture is well placed to advise on the design of software components and interfaces that will work well with any of these new data and information structures.

- **Business analysis:** can advise on the right approach to redesigning any affected business processes. Initially these include appointment booking, cancelling and rescheduling appointments, cancelling and rescheduling clinics, waiting list management and management information reporting.

- **SMEs:** staff from each medical department that runs a clinic can provide details of contents of messages, special instructions, timing of sending messages and other factors affecting communication.

- **Data and information architecture:** can advise on the design of the message content, format and storage, and on the storage and access of new data items such as new types of contact details and communication preferences.

- **Technical specialists:** security architects or specialists can advise on the security of electronic communication and (in the future) management of contact details by patients.

6.1.8 Stakeholders involved in solution implementation and deployment

The main focus of activity by the solution architecture team and others during the solution architecture life cycle is the production and validation of a solution design. Once the design is complete, it is ready for implementation and deployment.

The point at which the solution architecture team hands over the design to those who will organise and coordinate the delivery of the solution to the business is the roadmap development phase of the life cycle. The delivery roadmap is built with input from:

- **Solution architecture:** providing details of the change work that needs to be implemented.

- **Business stakeholders:** specifying priorities for the delivery of benefits.

- **P3M:** advising on the type and level of planning required to deliver the change work, and any dependencies between change activities.

Other stakeholders involved in implementation and deployment are:

- Solution development team.

- Business change.

- Service operations and DevOps.

The concerns of stakeholders involved in solution implementation and deployment include:

- Availability of skilled practitioners.
- Capacity of the business for change.
- Management overhead is minimised.
- Deliverables clear and explicit.
- The need for high-quality requirements.
- Minimal change to roadmap during implementation.
- Good communications between teams and practitioners.

Fallowdale Hospital patient communications: stakeholders for implementation and deployment

The following groups will potentially be involved in the implementation and deployment of the patient communication solution:

- **P3M:** there is a programme management office (PMO) that works with the business to put together project and programme teams and manage communication and documentation. Business change is within the remit of P3M.
- **Solution development:** this is carried out internally by staff in the IT department. This has traditionally been software development and maintenance but several projects have required more focus on data analytics, so there is now a team of data scientists.
- **Business analysis:** there are business analysts employed by IT as well as some within various lines of business. The former are assigned to solution development projects and are responsible for designing and documenting processes. The latter act as SMEs for the area of the business they represent.
- **Service operations and DevOps:** the hospital has a traditional service operations department that is part of IT. They are in transition towards DevOps and are looking at moving certain systems to be cloud-based, but the majority of systems are on premises or at a data centre. The patient administration system is managed externally at a dedicated data centre run by the supplier.

6.1.9 Service and product suppliers

Several component types that may form part of a solution are available from external service and product suppliers. These include:

- outsourced business functions;
- hardware and software;

- data and information sources;
- competencies via recruitment of new staff or training of existing staff.

Another category of supplier is that of business partners who supply components of a value chain, such as raw materials, packaging or other components that are part of a business service or product provided by the business. Since a solution is very likely to be part of a value chain, it can be hard to draw a distinction between these two categories, and there is certainly some overlap between them.

The concerns of all suppliers of services and products to the organisation or enterprise include:

- Reputation is enhanced or reinforced.
- Supply capability is broadened.
- Strategic aims are met.
- Profitability is increased.
- Relationship to procuring organisation or enterprise is strengthened.

Fallowdale Hospital patient communications: service and product supplier stakeholders

The hospital has many suppliers of services and products. The ones that are relevant to the patient communication solution are:

- The patient administration and clinic administration systems are provided by a healthcare software services company.
- Letter printer and folder is provided by an external company under a lease purchase contract. This includes the software used to manage the printing of letters using templates and enclosures. Day-to-day management of the system is performed by print room staff but the contract includes periodic servicing and urgent call-outs.
- Paper and other supplies for the letter printer and folder are obtained from various suppliers and organised by the hospital's procurement department.
- The postal service collects and delivers patient communication letters.
- The mainly internal email system is provided by the hospital IT department and this communicates externally via an internet gateway that is provided by an internet service provider.

6.1.10 Regulators and industry bodies

Many businesses are regulated by one or more outside regulators. For example, most jurisdictions have financial regulators. This goes beyond the normal requirement to obey laws or follow guidelines by, for example, imposing reporting and audit requirements and carrying out inspections. Breaches are punished in many ways, including bad publicity, fines or suspension of licences.

Some business areas have industry bodies that impose requirements on organisations and enterprises. These usually involve less direct regulation but rather rely on information received, for example from customers. Punishments are usually less severe.

Such external bodies are stakeholders for solution architecture and can be a source of requirements, constraints and concerns that must be accounted for in the solution design.

Fallowdale Hospital patient communications: regulatory stakeholders

The hospital receives public funding to provide healthcare services. A national government healthcare organisation sets national standards and receives reports summarising the care provided. Note that this organisation could also be categorised as an investor or even a customer of the hospital.

There is also a healthcare regulator, set up and funded by the government, which inspects and rates hospitals (and other care providers) and can insist upon remedial action in the case of failures.

Both these organisations have objectives to increase the fairness and equality of access to healthcare. This is a relevant concern for the patient communication solution, as good communication can enable and enhance fairness and equality of access.

6.2 STAKEHOLDER IDENTIFICATION AND EVALUATION

Stakeholders may be identified at any point in the solution architecture life cycle. It is common for stakeholder representation to change – for example, individuals may be replaced if they are required elsewhere in the business. This is regrettable as it can cause discontinuity in the relationship between solution architecture and the business. Ideally stakeholders should have a continuous involvement from the initiation phase, where the motivation for the solution is identified, through to the roadmap development phase, where the solution is handed over to those who will implement and deploy it.

Some stakeholders are listed in the architecture initiation document that is produced during the initiation phase. However, the main activity in the life cycle where stakeholders are identified and recorded is in the discovery phase and is called stakeholder engagement.

6.2.1 Stakeholder selection and allocation of roles

Accountability for the correct selection of stakeholders lies with the business owners and senior managers of the organisation or enterprise. They are also responsible for allocating roles such as business sponsor or product owner to individuals who have the authority and capability to fulfil them.

Business owners are a self-defining category and a very important stakeholder group. It is slightly less obvious who should be included in the associated category of decision makers: the senior managers. This includes people in posts from vice presidents (VPs) and C-suite members (CEO, CFO and others) to departmental managers.

Some senior managers have broad responsibility within the organisation or enterprise and will therefore have an interest in and concerns about the majority of proposed solutions. Others will have more focused responsibilities and be in charge of a narrower or more specialised area of the business. The latter group will normally only be stakeholders for solutions that fall within their area of responsibility.

A key decision is the allocation of the role of business sponsor or product owner. This is the person who authorises work to be carried out and signs it off when completed. They can make and overrule design decisions, although these are normally delegated.

Other roles that may be individually allocated include:

- Lead business analyst to operate as a bridge to the business area or areas covered by the solution.
- Technical coordinator to coordinate IT and security aspects of the solution. This is a standard role in the dynamic systems development method (DSDM) Agile methodology (DSDM Agile Business Consortium, 2014).
- Business visionary to represent business strategy and direction. This is a DSDM role that is appropriate for solutions involving large-scale change; for solutions with more limited scope, this role is integrated with the business sponsor.
- Liaison with external organisations, such as regulators and business partners.

6.2.2 Administration of stakeholder engagement

Responsibility for the administration of stakeholder involvement belongs to the solution architecture team. This includes recording stakeholders in a register, together with their concerns that will be used to validate the solution design.

The stakeholder register contains the following data:

- Name and ID.
- Role in organisation or enterprise.
- Contact details (of representative individual in the case of a group).
- Role in solution architecture, design and development including responsibilities, accountability and areas to be consulted on.
- Type or classification.
- Communication needs.

- Power and interest assessment (used to create a power-interest grid).
- Benefits sought from the solution.

Concerns are recorded in a separate register and linked to stakeholders with a many-to-many relationship, as a concern may be shared by multiple stakeholders and a stakeholder may have multiple concerns.

The concern register contains the following details:

- name and ID;
- details;
- examples;
- measurements and KPIs;
- plans for addressing;
- related viewpoints;
- governance.

The solution architecture team spends some time putting together the register based on artefacts from earlier activities in the life cycle and other inputs, and by consultation with the senior stakeholders that have already been identified.

Solution architecture is then responsible for communicating with each stakeholder in the register to confirm their role and elicit a complete list of their concerns at this point in the life cycle. Stakeholders may, of course, raise additional concerns at any point. Any significant changes, for example as a result of self-deselection, need to be referred back to the business sponsor or product owner for approval, or to other senior stakeholders if no one has been appointed to that role.

Thereafter, the solution team is responsible for enabling communication with and between stakeholders and for managing stakeholder interactions throughout the life cycle.

6.2.3 Inputs to stakeholder identification

The solution architecture team obtains the names and details of stakeholders from a variety of sources, including:

- Nominations by other stakeholders or others involved in problem analysis or solution definition such as P3M.
- Architecture initiation document produced during the initiation phase.
- Architecture artefacts catalogue constructed during the architecture inputs activity in the discovery phase.
- Enterprise stakeholder list from the enterprise architecture repository or continuum.

- Previous solution stakeholder lists.
- Organisational charts from the business.

6.3 MANAGING STAKEHOLDER CONCERNS WITH VIEWPOINTS AND VIEWS

Whenever concerns are identified, during the discovery phase or later in the solution architecture life cycle, they must be recorded in the concern register and linked to one or more stakeholders. The solution architecture team is responsible for managing the concern register and communicating changes. The business sponsor is accountable for concerns being raised and addressed by the solution.

To manage stakeholder concerns, the solution architecture team needs to ensure the currency and accuracy of the concern register. This can be through active management where the contents are reviewed on a regular basis to ensure concerns are being managed in the most appropriate way. Concerns are ultimately addressed by solution design decisions and so they will be discussed during stakeholder interactions, such as update sessions, workshops and decision-making meetings, including those of design authorities.

A key technique for handling the stakeholders and their concerns is by defining viewpoints that focus on specific concerns.

6.3.1 Defining viewpoints

Viewpoints are used to demonstrate the behaviour of certain parts of an AD that are of particular interest (see Figure 6.1). A viewpoint can define the following:

- **Selection criteria:** a filter that defines what data and information is to be extracted from the AD.
- **Artefact structure:** a template for how the data and information is to be organised into artefacts.
- **Processing instructions:** a specification of how the information in the artefacts is to be presented for ease of understanding and possible manipulation by stakeholders.

Note that the visualisation aspect of viewpoint definition is optional.   **i**

The use of visualisation to present information helps to make it accessible to stakeholders. It can be used to simplify and clarify the contents of a view to make it relevant for a broader range of stakeholders. Visualisation tools can also allow information to be manipulated for further analysis, for example by further filtering, grouping and changing parameters.

Figure 6.1 Relationship between viewpoint, view and visualisation

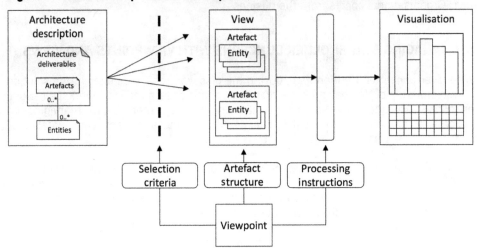

Fallowdale Hospital patient communications: viewpoint definition

One of the original issues that led to the need for a solution is the high cost of sending letters to patients.

The finance director (FD) of the hospital has expressed a concern that the new solution will not reduce that cost.

The solution architecture team has decided to create a viewpoint that focuses on the cost of communication so that views can be produced to address the FD's concern.

Following an interview with the FD, it has been proposed that the following elements be included in a viewpoint:

- Total cost of patient communications.
- Breakdown of cost by solution component, including IT support.
- Breakdown of cost by type of communication.
- Analysis to distinguish fixed and marginal costs.

Activity 6.1

The Fallowdale Hospital FD has identified four elements that can address the concern about whether costs will be reduced by the new solution. These elements are the basis for a stakeholder viewpoint.

Given the proposed solution architecture shown in Figure 6.2, which components are relevant for this viewpoint?

Figure 6.2 Proposed architecture for patient communication solution

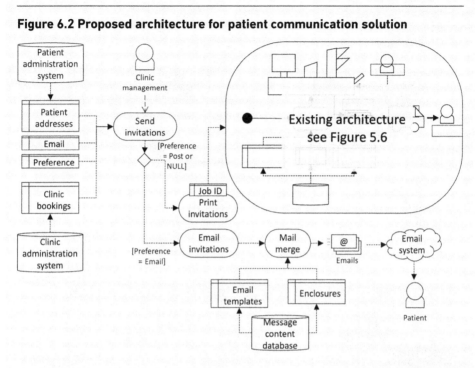

What type of visualisation would be appropriate to show the effect of the new solution on the costs of patient communication?

6.3.2 Producing views

A viewpoint can be used to produce many views. For example, an essential view is of the current or baseline architecture. Producing this view shows the existing situation and can be extremely helpful in highlighting an issue that solution architecture is trying to address.

It can also be used to check that the viewpoint includes all the information needed to address the concerns expressed by the stakeholders, and that the organisation and visualisation of the information is as useful as possible. Stakeholders can only really judge the efficacy of a viewpoint by looking at the views it produces, so it is a good idea to produce baseline views as a way of obtaining stakeholder feedback to refine the viewpoint.

Having established the current state of its area of concern, the next use of a viewpoint in solution architecture is to examine the situation that would be in place after a proposed solution has been deployed. Any view based on a proposed solution design is necessarily a forecast, so it cannot be verified in the same way as the equivalent baseline view, where actual facts and figures are available for comparison. A view based on a target

AD can be used to inform stakeholders about their concerns. If the situation projected by a viewpoint based on a target AD is not acceptable to stakeholders, the solution architecture team can redesign the solution to improve certain aspects and another view can be produced and communicated to stakeholders.

In some cases, particularly where large-scale change is planned to take place in a number of smaller increments, there may be multiple intermediate ADs as well as a final target AD. A view can be produced for each AD in this series. This can be used to show a progressive change in the area of concern that is the focus of the viewpoint.

Finally, after a change has been deployed and the new architecture documented in an AD, a view can be produced based on the actual situation and compared with the forecast, to check whether the expected benefit in the area of concern has been achieved.

In the solution architecture life cycle, the majority of views are produced during the validation phase. This is because a full AD for the proposed solution has been produced during the logical design phase. However, viewpoints can and should be created throughout the life cycle, whenever a new stakeholder concern is identified.

6.4 DEVELOPMENT AND SELECTION OF BUSINESS CASE OPTIONS

During the early phases of the solution architecture life cycle, some ideas may be put forward as to the form of the solution that should be designed, implemented and deployed. These may exist in the minds of those who first identify that there is a problem to be solved. Other stakeholders may have ideas upon the subject and will no doubt discuss these with others, especially during the discovery phase when requirements are being captured.

However, when it comes to preparing a small number of options to include in the business case for discussion and selection by all stakeholders, the business sponsor is the person with responsibility for deciding and formally nominating them.

6.4.1 Documenting business case options

The solution options that are presented to stakeholders for analysis, comparison and decision need to contain sufficient relevant information in a consistent form to allow the process to work effectively and for an informed decision to be made that is in the best interests of the business. Therefore the task of collating and documenting the options is the responsibility of the solution architecture team, which has access to all available information and is in communication with all the stakeholders.

6.4.2 Supporting information

Each business case option needs:

- description, ideally including a diagram;
- gap analysis to show the scope and scale of change;

- cost benefit analysis;
- return on investment forecast;
- assessment of the type and level of risks involved.

The solution architecture team is responsible for collating this information and making it available to all solution stakeholders to facilitate their decision making. The solution architecture team can ensure that the options are comparable, and that the information provided is produced to the same standard. This avoids risks such as supporters of a particular option only seeing the positive aspects, and this being reflected in the information provided for that option.

Some of the information needed will have been assembled by this point, but there is usually a need to obtain supporting information to fill in any gaps and for the purpose of validation and due diligence.

Enterprise architecture is responsible, upon request, for providing models of the current state that can be used, for example, in preparing gap analysis estimates and risk assessments of the options being proposed.

Enterprise architecture and its subdomain architectures may be consulted for details about the availability of existing components and the options for creating new components that have been identified as part of one or more of the business case options.

P3M may be consulted if any option requires outline planning, for example to establish what resources will be required or any risks involved.

6.4.3 Presenting options, facilitating discussion and decision making

The solution architecture team makes the documented business case options available to all stakeholders for consideration. Ideally this is done simultaneously to allow each option to receive equal consideration. If it is likely that amendments to the documentation will be requested by stakeholders, then some form of version control should be in place with any changes being notified to all stakeholders. This is more likely where there are many stakeholders involved.

Note that although not all stakeholders will have a say in selecting the preferred option, it is still essential that the information is made available to all of them. This allows for feedback and other useful input to be received that might better inform those who will make the decision.

A very common way of facilitating discussion of the options is by holding a workshop for those stakeholders who are in a position to influence the final decision. Stakeholders with less influence are often also included so that they can at least provide input and be represented. The business sponsor is accountable for the choice of the option that is best for the business or rejecting all the options if none are deemed suitable to provide the required solution. The workshop will be led and facilitated by the solution architecture team; this includes recording the discussion and the decisions made and making the documentation available to all stakeholders afterwards.

6.5 SCOPE DEFINITION

The scope defines what is included in the solution and also what is excluded from it. A solution is made up of a number of components and the interfaces between them, so the scope can be delineated in a number of ways to include:

- business services and products;
- business functions;
- geography;
- data and information sets;
- information systems;
- business aims and objectives.

Components (and their interfaces) that are in scope can be:

- **Part of the problem:** existing components that cause or contribute to the problem being addressed.
- **Part of the current solution:** existing components of a current solution in the problem area.
- **Not currently part of the solution:** existing components that are used elsewhere.
- **Required but not yet available:** components that are required (according to the selected business case option) to solve the problem but do not currently exist within the organisation or enterprise.
- **Impacted by the solution:** components that are not part of the existing or proposed solution but may be impacted by it, either positively or negatively.

The business sponsor is accountable for the scope matching the needs of the business. Other stakeholders are involved in identifying and verifying the components and interfaces that are in scope for the solution.

6.5.1 Business components

Business owners and senior managers are accountable for the correct identification and verification of changes to business services, processes, organisational structures and job roles that are required to fully enable the benefits of the selected solution. However, the responsibility for identifying business components that are in scope for a solution is usually delegated to operational (line of business) managers, business architects and business analysts who have more detailed knowledge of their functions and interactions.

Fallowdale Hospital patient communications: business components

The operations director is in overall charge of the print room activities, which encompass a number of business (as well as technology) components that are included in the baseline solution for patient communication. The operations director has delegated responsibility for identifying and verifying changes to business components as well as reporting back on potential efficiencies that can be achieved as part of the change process. The operations director retains accountability as the senior manager in charge of the business area.

Activity 6.2

Apart from the print room, what other business components are in scope for the Fallowdale Hospital patient communications solution? Which stakeholders can validate this by providing the necessary knowledge about their functions and interactions?

6.5.2 Information and technology components

Various stakeholder groups are consulted to establish the complete list of information, data and technology components that are in scope for the solution.

For example, information and data components are found within data architecture. Technology components may be found within applications, software or technology architecture.

Fallowdale Hospital patient communications: information and technology components

The applications architecture function has identified some existing components, some of which have been used previously at the hospital as part of externally provided applications. They provide functionality that validates email addresses and other contact details.

One component validates patient postal addresses with a national healthcare demographic database and verifies postal codes using the postal service via a REST API.

Other components validate email addresses using either the national healthcare demographic database, commercial provider and social media identities, or a central government notification service.

These options have been suggested to the solution architecture team so that they can be considered for use within the patient communications solution.

6.5.3 ABBs and SBBs

Enterprise architecture and its subdomain architecture functions have an overview that covers the entire organisation or enterprise. These functions can therefore check and confirm the components that have been identified as part of the scope and identify any that have potentially been missed.

These can include:

- **ABBs:** such as template designs and patterns that can be used within a solution.
- **SBBs:** component configurations that have been used as part of a previous or existing solution.

Fallowdale Hospital patient communications: architecture and solution building blocks

The enterprise architecture function has identified a building block that is part of an existing solution. This is part of the employee management solution and allows clinicians and other staff to maintain their own contact details.

This building block has been proposed to the solution architecture team as a possible component to allow patients to manage their own contact details.

At this stage it would be an ABB, as it will be part of the architecture and could be used as a design pattern for managing this type of information. It could become an SBB if the components and their configuration from the previous solution are replicated in this one.

6.5.4 Documentation and sign-off

The scope is documented in the form of a list of components that are in scope, indicating whether they exist or not, and links to further information. This list may be supported by models explaining the role played by the components in the existing or proposed solution.

All stakeholders are informed that the scope is complete and are given the opportunity to review the documentation and provide feedback.

Once any feedback has been reviewed and taken into account and any amendments made to the scope documentation, the business sponsor signs it off as accurate and complete.

6.6 VALIDATION AND SIGN-OFF OF CHOSEN SOLUTION

The point of identifying and involving stakeholders throughout the solution architecture life cycle is that the solution that is designed is the best one for the business as a whole. This may involve balancing competing interests among the stakeholders, but the design that enters the validation phase should be recognised by all stakeholders as the ideal solution for the problem, all circumstances considered. There should certainly be no surprises, providing the stakeholders have been involved appropriately up to this point.

6.6.1 Concerns dealt with

For stakeholder concerns to be dealt with appropriately, they must first be identified. Once identified, they are documented, linked to the stakeholders that hold them and managed through the life cycle. This approach aims to ensure that they are part of the process throughout and are taken into account at every stage of design.

All stakeholders are accountable for ensuring that all their concerns about a problem area or proposed solution have been identified and documented. Identification can take place at any point in the life cycle, although earlier is better, as this means they are more likely to be part of the design rather than requiring a revision after design work has been done.

Stakeholders also need to be satisfied that their concerns are addressed by the design. This can be done by designing viewpoints and producing views from them. This technique can be used at any point in the life cycle, but is most effective during the validation phase when the target AD of the solution is complete.

It is possible that a view will highlight an effect of the design that is unacceptable to one or more stakeholders. In this situation there are two options:

- **Solution redesign:** remove or mitigate the unacceptable aspect.
- **Accept the effect:** leave the design unchanged, but be aware of the impact and take appropriate operation decisions when the solution is deployed.

If this happens, it is usual to have a broad discussion among stakeholders or to refer the issue to a design authority or other meeting so that it can be put in context. It is frequently the case that stakeholder interests have to be balanced and this should be done irrespective of the type of decision that will be made.

6.6.2 Design thoroughly tested

A solution design is subject to testing by the solution architecture team throughout its development. Each design element that is introduced aims to satisfy or fulfil one or more requirements. The design is tested by the solution architecture team, first to see if the new requirement has been met and subsequently to check that previously satisfied requirements have not been impacted (regression testing).

Testing can be used as an opportunity to involve other stakeholders. This can be done individually, where the solution architecture team shares an aspect of design that is pertinent to a particular stakeholder, or in a workshop, when several design decisions can be discussed.

During the validation phase there is an activity to test against enterprise architecture models. This activity works in support of the use of viewpoints and views, which focus on specific stakeholder concerns. Testing against EA models allows the solution architecture team to check that the solution works as anticipated in a broader context.

The solution architecture team is responsible for testing the solution against current state EA models and demonstrating, where necessary, that the desired benefits of the solution will be achieved, the business requirements met, and no constraints broken.

The solution development team, including business analysis and software architecture, may be consulted and have responsibility delegated for testing the solution design and feeding back any concerns relevant to those disciplines.

Infrastructure architecture is responsible for performing the solution technology definition that can occur provisionally throughout the solution architecture life cycle. Infrastructure architecture can use a similar process to test that the solution design will work as planned within the existing architecture landscape, or that any additions or modifications to the infrastructure will not break any constraints.

Testing may be performed by the solution architecture team, either working alone or in consort with other stakeholders. For complex solutions with large numbers of requirements or involving significant business change, it can be worth involving a team of independent testers to prepare and execute a formal testing programme.

An effective technique for testing with business stakeholders is to design **business scenarios** that exercise the solution requirements. These work to demonstrate success in terms that are relevant and clear to business stakeholders.

6.7 GOVERNANCE AND DESIGN AUTHORITY

All stakeholders may play a part in the governance of the solution design, implementation and deployment. Some stakeholders have specific governance roles:

- **Business sponsor:** accountable for all decisions made throughout the governance process.
- **Solution architecture:** responsible for constituting design authorities and managing the decision-making process, including meetings, circulation of reports, collating and communicating feedback, and ensuring that all stakeholders are fully informed about governance activities.

Other stakeholders are involved in governance by being members of one or more design authorities.

Since solution architecture is responsible for facilitating appropriate communication with and between stakeholders, it is appropriate to create and constitute a design authority for the solution (see Figure 6.3). This is an excellent way to provide structure for the large amount of communication required to ensure the engagement and involvement of a large number of stakeholders in the design, implementation and deployment of a solution so that it is aligned with the business.

A design authority is constituted like a formal committee with specific roles and terms of reference to define what decisions it must make and the process for doing so, including escalation to the business sponsor and referral to solution architecture.

The use of design authorities is a way of facilitating discussion and promoting understanding among stakeholders. Reports from design authority meetings and workshops are made available to all stakeholders with key points highlighted. This promotes understanding of the design decisions throughout the stakeholder community.

Activity 6.3

The Fallowdale Hospital board has identified the need for a design authority for the patient communications solution. They want to limit the number of people to six or fewer, but it needs to be representative of all stakeholders and able to cover all aspects of the solution. The commitment to attending meetings and reviewing documentation is not expected to be very demanding, but the board is mindful of the need to turn around decisions quickly so as to avoid becoming a bottleneck.

Identify up to six stakeholders who could make up the solution design authority and provide an explanation for this choice.

Design authorities consult with specialists where their expertise is required and does not exist within the design authority. Specific input or general consultation may be requested, and this may be given by invitation to present at a meeting or workshop or in a written report.

For complex solutions, more than one design authority may be required, each focusing on a different aspect of the solution. Design authority members are selected both for their specialist knowledge of the business and to provide breadth of coverage from all stakeholder groups.

Design authority members are responsible for making decisions and recommendations about the design, implementation and deployment of the solution.

The business sponsor must be consulted to ensure that the remit and escalation paths of design authorities are correctly stated in their terms of reference so that they can make decisions at the appropriate level.

Figure 6.3 Design authority

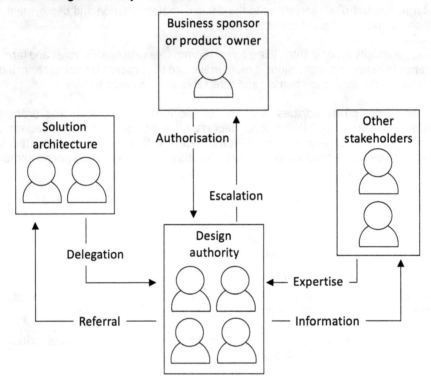

Activity 6.4

The agenda for the first meeting of the Fallowdale Hospital patient communications solution design authority includes the following items:

- Suggested changes to the letter printer and folder if there is a reduction in capacity.
- Selection of appropriate safeguards to preserve the privacy of patient information.
- The best approach to getting patient consent for the various communication methods (email, text and letter).
- How to control the content of messages that are sent out on behalf of the hospital.

Given this agenda, might the design authority need any expertise from other stakeholders or specialists to support their decision making, and if so, who might they call upon to attend the meeting to give their input?

REVIEW QUESTIONS

1. Which two of the following are required to produce a view in solution architecture?

   i. Architecture description (AD).

   ii. View processing system.

   iii. Business rule.

   iv. Viewpoint.

   v. Visualisation.

   a. i and ii only.
   b. iii and iv only.
   c. ii and v only.
   d. i and iv only.

2. What term is used for the person who is the public face of a solution and has the necessary authority to ensure its success?

   a. Business owner.
   b. Business architect.
   c. Business sponsor.
   d. Solution architect.

3. What type of communication between stakeholders can take place **without** involving the solution architecture team?

   a. Expert advice to a design authority.
   b. Selection of a business case option.
   c. Notification that the scope has been signed off.
   d. Stakeholder selection and the allocation of roles.

4. The solution architecture team is testing a solution design and has asked for a business analyst to assist in walking through the solution. What is a possible reason for asking a business analyst to help in this way?

   a. Solution testing is a core competency of a professional business analyst.
   b. Business analysts can identify design issues concerned with the interface between business processes and other components.
   c. The business analyst can re-engineer the business processes.
   d. Business analysts have the best overall understanding of the concerns of the business.

7 SOLUTION TECHNOLOGY DEFINITION

LEARNING OUTCOMES

When you have completed this chapter, you should be able to demonstrate an understanding of the following:

- The purpose of solution technology definition and the steps involved
- Precursors, requirements and context
- Baseline opportunities and target constraints
- Mapping solution building blocks to infrastructure services
- Networking infrastructure
- Achieving functional requirements and NFRs
- Assuring end-to-end security
- Identifying gaps in infrastructure services

7.1 THE STEPS IN SOLUTION TECHNOLOGY DEFINITION

The aim of solution technology definition is to show:

- What type and level of technical infrastructure is required to deploy a solution.
- What areas of the current infrastructure will need to change.
- How much cost, time and other resources will be required to make the necessary changes.

Solution technology definition is based on a model that allows solution components (SBBs) and interfaces between them (SBBIs) to be mapped to infrastructure components using a step-by-step approach (see Figure 7.1).

Each component has a set of functional requirements and NFRs and some, but not all, of these requirements will map to infrastructure service requirements. An infrastructure service requirement must include the level of service required.

Figure 7.1 Solution technology definition model

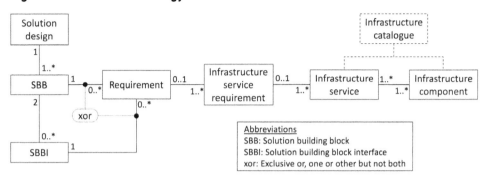

Infrastructure service requirements map to suitable infrastructure services that can meet the requirement. If an infrastructure service exists, or is planned for, it will appear in the infrastructure catalogue and will be mapped to one or more infrastructure components. If a service requirement does not map to a suitable infrastructure service, this indicates a gap.

Solution technology definition is a critical activity in making the transition from logical design to physical implementation and ultimately deployment. The key **stakeholders** involved in this transition are:

- The **solution architecture team** who have produced and validated the logical design.

- The **infrastructure architecture function** who will fit the physical technology platforms and enabling infrastructure that support the solution into the infrastructure architecture and model any necessary changes.

The steps outlined here are arranged in a formal process that can be used to produce a solution technology definition using the outputs from gap analysis. A less formal process is often used during solution design to assess the physical dimension of a solution. For example, as soon as any infrastructure service requirements can be identified, the solution architecture team can discuss how they might be implemented in the current infrastructure architecture landscape and how they might fit into strategic planning.

Fallowdale Hospital patient communications: early stage solution technology definition

During the solution outline definition phase, it has become clear that the capability to send emails would be in scope for the solution. The solution architecture team has decided to consult with the infrastructure architecture function to understand any technological aspects that might affect the design.

Infrastructure architecture has raised the following points:

- There is an existing email system, mainly for internal communication but occasionally used for external contacts, but with a ban on patient details being sent or received.

- Existing provision would need to be scaled up or a new system provided.

- There is a national email system that is highly secure and can be used to send patient details, but it can only be used between healthcare professionals.

- The IT strategy contains an objective to transfer infrastructure components to cloud hosting, and this includes the current email system.

- A cloud-based email system would have advantages for scalability and other NFRs, result in lower costs, and mean internal staff could be included as recipients.

A provisional decision to use a cloud-based email system has been made and communicated to stakeholders with an outline of the benefits of this approach.

The steps to define the technology for a solution are:

- **Clarify precursors, requirements and context**, ensuring all inputs to the process are prepared and disclosed to all parties, and that any initial queries are addressed before the process starts.

- **Understand baseline opportunities and target constraints**, mapping the scope of the solution to the baseline infrastructure architecture, and identifying any potential opportunities and constraints within the scope.

- **Map SBBs to infrastructure services**, itemising infrastructure services and components that meet the solution requirements.

- **Define networking infrastructure**, modelling the connectivity required to fulfil the needs of the interfaces between solution and infrastructure components.

- **Iteratively refine to meet functional requirements and NFRs**, testing the target infrastructure architecture against the solution requirements and making modifications where necessary.

- **Assure end-to-end security**, assessing the solution and infrastructure architectures against security standards and removing or mitigating vulnerabilities.

- **Identify gaps in infrastructure services**, itemising any gaps in provision or capacity and reporting on the costs and other implications for change.

- **Validate the solution technology definition**, reviewing with senior stakeholders and obtaining sign-off from the business.

7.2 PRECURSORS, REQUIREMENTS AND CONTEXT

The starting point for solution technology definition is the solution design. This contains the components and interfaces that make up the solution and define its scope.

Any technology needed for the solution must be within this scope. For both solution and infrastructure architecture, however, impacts may extend beyond the scope of the solution. For example, a change to an existing system could impact other users even though they are out of scope for the solution.

Other models, artefacts and documents can provide additional detail to support the process of solution technology definition. The solution architecture team is responsible for making all necessary information available for use by the infrastructure architecture function and any other roles involved. This can include:

- logical solution design;
- baseline and target solution architecture descriptions;
- gap report and associated gap models;
- business functional requirements and NFRs;
- viewpoints and views that address infrastructure-related areas of the solution.

The infrastructure architecture function is responsible for reviewing the solution design and associated information and raising any queries. Queries typically concern clarification of design decisions, perceived gaps in the information provided, or projected timescales that could affect the availability of infrastructure components.

The solution architecture team is responsible for ensuring effective communication with stakeholders at all times during the solution architecture life cycle. The solution architecture team is accountable for the fact that all stakeholders are informed about all activities and decisions that affect the solution. In addition they are accountable for the fact that stakeholders who are directly affected by a decision or who have registered a related concern are consulted before the decision is finalised. This includes escalating any significant decisions, such as those that may impact on the business requirements, budget or timescales, to the business sponsor for approval.

7.3 BASELINE OPPORTUNITIES AND TARGET CONSTRAINTS

Baseline opportunities are beneficial changes that can be made to the baseline infrastructure architecture as part of the implementation of a solution. These are opportunities to incorporate changes to the infrastructure architecture with those required for the solution. This approach can save effort, reduce costs and bring forward desired changes that would normally require a separate change project or programme.

Note that baseline opportunities can occur in other subdomains of enterprise architecture such as business, data, information or applications architecture. For example, the implementation of a solution may provide an opportunity to rationalise or redesign business processes that would otherwise have had to wait for a separate programme of work to be authorised. However, the opportunities being identified and considered in solution technology definition are primarily from the infrastructure architecture domain.

A major source of opportunities is the IT strategy. This represents decisions that have been made by business leaders about the future direction of IT infrastructure within the organisation or enterprise.

Fallowdale Hospital patient communications: baseline opportunities

The Fallowdale Hospital IT strategy contains the following items:

- **Automation:** automate any process where the rules can be defined. A robotic process automation (RPA) scheduling tool has been trialled for booking operating theatre time.

- **Communications:** improve equality of access to healthcare by using technology to provide rapid responses to patient queries about their care. One component that is under trial looking for a low-risk project is a chatbot that can interact with patients in their own language.

- **Cloud:** reducing IT costs and making the best use of IT assets is a priority. New IT systems and services need to be assessed for cloud hosting as this has a lower capital and running cost. Software development and testing must be done in cloud environments and use open source code where possible.

- **Data:** maximise the use of information to improve the service provided to patients, base all decisions on the best available data, use analytics to plan ahead, use machine learning and artificial intelligence where it provides benefits, use flexible data storage technology to capture data with emerging requirements and to more closely match business requirements.

Activity 7.1

Based on the hospital's IT strategy, identify one or two baseline opportunities where existing strategic aims can be achieved by choices made during the solution technology definition process.

The other area that can be revealed by a close inspection of the overlap between infrastructure architecture and solution design is that of **constraints**. Constraints are limits within which infrastructure and solution architecture must work to define the technology that will be part of the implementation of the solution. These constraints also apply to those who will implement the solution in the completion phase.

There are many different types of technological constraint. The main ones are shown here:

- **Approval or preference:** a list or model of technological choices that have been approved for use within the organisation or enterprise, placing a constraint on infrastructure components that can be used as part of a solution.

- **Standards:** infrastructure architecture contains standards that must be adhered to and therefore can constrain a solution; an example is the ISO 50001 standard for energy management, which aims to reduce energy footprint and greenhouse emissions (ISO 50001:2018, 2018).

- **Budget:** financial limits are placed on the overall spend on IT and priorities established for how the money should be spent, putting a constraint on the changes that can be made to the infrastructure to accommodate a new solution.

- **Time:** some required technology may not be available in time for the solution to take advantage of its benefits, or there may be pressure to stop using certain technologies by a certain date.

- **Contracts:** existing contracts must be adhered to and this can impose constraints on what can be changed.

- **Geography:** infrastructure services may not be universally available for various reasons and could result in different components being used in different geographical locations, especially if governments place restrictions on their use.

- **Skills:** the skills of staff to implement, deploy or manage its components may be a constraint on certain aspects of a solution, where some development tools or environments may not be available to be part of a solution.

The idea of approval or preference of certain infrastructure technologies, or products that implement them, has been adopted by many organisations. This has a number of benefits for infrastructure architecture. It acts to limit the number of different technologies in use by controlling the choices available. It can also help by ensuring that technologies and products have been through a process of assessment and trial.

A simple way for an organisation or enterprise to express technology preferences is to use the infrastructure catalogue. Each item in the catalogue requires a status to indicate the type of preference expressed. This can be complicated or simple. A simple system involves a RAG or traffic-light rating, so that each item is either red, amber or green. Here, red indicates it is not to be used for new projects and any opportunity to remove it from the infrastructure should be acted upon. Green indicates that this is a preferred technology or product and should be the first choice. Amber items may still be used, but approval to do so by the infrastructure architecture function or another authority is required.

An alternative, more dynamic approach is a technology radar that gives technologies and products more of a life cycle.

i

Technology radar

A useful analogy for the rapidly changing nature of technology is provided by a radar system such as those used at airports to track the flight path of incoming aircraft.

Technology also has a kind of flight path and IT professionals even speak of their awareness of a particular emerging technology as being 'on the radar'. This is used in a formal way by many organisations to keep track of their interactions with technology. It can then serve as a guide to the suitability of different technologies as part of the infrastructure architecture and therefore as components of a solution.

The technology radar shown in Figure 7.2 would be accompanied by a key showing what technology is represented by each of the items shown on the radar. A technology can only be in one segment and one ring at any point in time, although it is possible for one version to be in **deploy** and a newer version to be in **trial**.

This can be as simple as categorising technologies, and products that implement them, as:

- **Trial:** suitable for a trial ahead of being accepted for use.
- **Deploy:** ready for use.
- **Decommission:** replace as soon as possible.

Other levels are sometimes used, such as having an assessment stage before trial or having a retirement stage before decommissioning. For example, an additional process of authorisation could be required before technologies outside of the 'Deploy' ring can be used as part of a solution.

Like an airport radar, a technology radar needs someone scanning the horizon for new technological developments and making decisions about their relevance to the organisation or enterprise. This leading-edge awareness of innovation needs to be backed up by processes and procedures to ensure that the most useful and relevant technologies are used to benefit the business.

One important benefit of using a technology radar is that it communicates enterprise decisions to a wide audience in a clear and accessible way. It is extremely useful for stakeholders to be aware of what is available for use or undergoing a trial.

The process for identifying and understanding opportunities and constraints takes place between the solution architecture team and the infrastructure architecture function. Other stakeholders may be consulted to support decisions about the effect on the solution design. There are also levels of decision that require escalation to senior business stakeholders, including the business sponsor.

A RACI matrix (see Table 7.1) may be used to show which stakeholders have the responsibility (R) and accountability (A) for carrying out activities, together with those who may be consulted (C) and also those who must be informed (I) about decisions and other outcomes.

Figure 7.2 Technology radar (actual technology names are not shown)

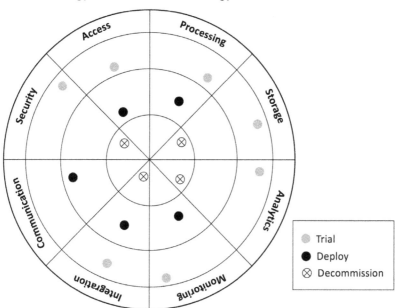

Table 7.1 RACI matrix: Understanding baseline opportunities and target constraints

Activity	Infrastructure architecture	Solution architecture	Enterprise architecture	Business sponsor	Business owners and senior managers	All stakeholders
Present baseline infrastructure architecture	R	I				
Highlight overlap with solution	RA	C				
Review baseline infrastructure architecture		R				
Clarify and query to fully understand baseline infrastructure architecture	C	RA				
Identify and document opportunities and constraints	RA	CI	C			I
Document cost, risk and other impacts on infrastructure architecture	RA	I	C	C	I	C
Identify and document impacts on solution design	I	RA		I		I

The enterprise architecture function must be consulted if any major change to the infrastructure architecture is proposed to support the implementation of a solution. This is to ensure that any change is consistent with enterprise principles and policies.

Any significant impact on capital or recurring costs must be escalated to business owners and senior managers and their approval sought.

7.4 MAPPING SBBS TO INFRASTRUCTURE SERVICES

SBBs that map to infrastructure services may be of different types:

- **Technology:** equipment or software.
- **Process:** business process or procedure.
- **Information:** data structure, message, document or other unit of information.
- **Organisation unit:** team, department, organisation or other business unit.
- **Person or role:** business actor or end user of a service.

This activity depends on having a list of SBBs with requirements that can be met by infrastructure services. Where there is a baseline solution architecture that is being replaced with a modified solution design, it would be reasonable to remove any components that do not change from this list. However, this approach risks missing some potential benefits that could be achieved by mapping an existing SBB to an infrastructure service.

Fallowdale Hospital patient communications: including existing, unchanging SBBs when mapping to infrastructure services

The patient communication solution has these two SBBs:

- **Letter printer and folder:** existing component, unlikely to change apart from a reduction in volume.
- **Email generator:** new component that produces emails containing the same information as letters but as an alternative way of delivering the information to patients.

The email generator has a requirement to log its activity so that there is a record of communication with patients.

The letter printer and folder does not have such a requirement because its software already has a log that was considered sufficient when it was first installed.

When this aspect is examined during solution technology definition, a logging service is selected as it:

- can keep a reliable record of transactions;
- records all necessary information;
- is backed up and archived in line with enterprise architecture directives;
- can be integrated with patient records for a 'single view of the customer'.

There is a decision to be made about whether the letter printer and folder should also use the logging service. If it had been excluded from the list of SBBs under consideration, then it is likely that this opportunity would have been missed.

Table 7.2 shows a typical infrastructure service catalogue with services categorised by function. They can additionally be mapped to infrastructure components in an infrastructure technology catalogue.

Table 7.2 Infrastructure services catalogue

Category	Service	Category	Service
Client access	Devices	**Content delivery**	Content delivery network (CDN)
	Desktop management		
	Remote desktop		
	Virtual client		
	Website		
	Landing zone		
Processing	Application server	**Communication**	Email
	Data centre		Instant messaging
	Serverless computing		Text
	Containers and management		Voice
			Pager
	Machine learning		Video
	Artificial intelligence		Letter printer
			Postal system
Data storage	Database	**Security**	Antivirus, anti-malware
	Big data/NoSQL		Firewall
	Backup		Encryption
	Archive		Identity and access management (IAM)
	Logging		

(Continued)

Table 7.2 (Continued)

Category	Service	Category	Service
	Document management		Certificates
	Secure filing cabinet		Smartcards
			Remote access devices
			Auditing
			Logging
Data analytics	Data lake	**IoT**	Sensors
	Management Information		Mobile equipment
	Business Intelligence		Transport management (vehicles)
	Data warehouse		
	Dashboard		
Enterprise applications	Payment	**Printing**	Document printing
	CRM		Plotting
	Office productivity		High resolution printing
	Finance and accounting		Barcode labels
	Product management		
	Human resource management		
	Resource planning		
Monitoring	Application logging	**Scanning**	OCR
	Log analysis		OMR
			Image
Application integration	API (separate catalogue)	**Development**	Environments (build, test, train etc.)
	Integration engine/ service bus		Integrated development environments
	Business rules engine		Continuous integration and delivery (CI/CD)
	Data conversion		
Networking	WiFi	**Location/GIS**	RFID/NFC
	Wired access		Barcodes
	High-speed networking		GPS (vehicles)
	Internet access		
	VPN		

Activity 7.2

Fallowdale Hospital has identified the need to run as many clinic appointments as possible remotely with the patient at home or another healthcare setting, such as a care home or GP surgery.

Identify any additional infrastructure services that may be required to enable this capability. Give a brief explanation of why each service will be required and the impact it will have on the patient communications solution.

The process for mapping SBBs to infrastructure services is driven by the infrastructure function in consultation with the solution architecture team. Decisions may be escalated if they go beyond the remit of infrastructure and solution architecture. Specialist knowledge may also be required, and other stakeholders can be consulted if necessary.

A RACI matrix (see Table 7.3) may be used to show stakeholder involvement in this activity.

Table 7.3 RACI matrix: Mapping SBBs to infrastructure services

Activity	Infrastructure architecture	Solution architecture	Business sponsor	Enterprise architecture	End users and business actors	Data or information architecture	Security architecture and specialists
Itemise SBBs and their infrastructure service requirements	RA	CI				C	C
Categorise infrastructure services as existing, requiring modification or new	RA	C					
Validate that all business requirements are fulfilled by the identified infrastructure services	C	RA	C				
Validate that all enterprise directives have been followed	RA	I		CI			C
Validate that the minimum level of change, cost and disruption has been achieved	I	RA	I	C	C	C	C
Authorise capital and recurring costs for the provision of new or modified infrastructure services	CI	CI	RA	CI			

For solution requirements where there is an existing infrastructure service provision, reconfiguration may be required, or the capacity expanded. These details need to be recorded for each SBB.

Fallowdale Hospital patient communications: mapping SBBs and their requirements to infrastructure services

The patient communications solution contains the following SBBs, which are shown with some of their requirements:

- **Communications management system:** manage patient preferences, report exceptions, initiate communication based on trigger, manage responses (in future iterations), record activities.

- **Email generator:** produce and send emails, correct person and appointment details, maximum 30 minutes from receipt of instruction, keep record of each email sent, volume estimated to start at 100 per day, rising to 700 per day within a year.

- **Letter printer:** print letters, correct person and appointment details, maximum 30 minutes from receipt of instruction, keep record of each letter sent, 90 per cent of letters ready for collection at 4.00 p.m., volume currently 800 per day and estimated to fall to 100 per day in eight months.

- **Manage waiting list:** transfer patient from waiting list and allocate to a clinic appointment slot, slot must be appropriate to patient condition and stage of referral or treatment, escalate if no slot found within the time limit from referral.

- **Medical or diagnostic unit:** manage clinics to meet demand, ensure accurate clinic details, ensure instructions are complete and correct.

- **Message contents:** clear communication that can be understood by recipient, accurate contact and appointment details, complete instructions to enable attendance, relevant hospital messages, minimum personal data.

- **Patient communication preferences:** will become recipient communication preferences in future, low confidentiality, available on demand, currency confirmed within 12 months, flexible with potentially unlimited forms of communication.

- **Patient record:** specifically contact details, available on demand, currency confirmed within 12 months, 100 per cent compliance with data model, address verified, email format validated, highly confidential.

- **Post room:** monitor printing, refill supplies, escalate service calls, fill post bags, securely transfer letters to postal service.

- **Reschedule appointment:** verify request, find new slot or move to waiting list, repeat for any linked appointments, notify department of any newly free slots.

Activity 7.3

Select up to five of the Fallowdale Hospital patient communications SBBs listed above and:

    a.   classify each as either technology, information, process, person or role, or organisation unit;

    b.   identify any requirements that can be mapped to infrastructure services, and specify the type of infrastructure service that could support the requirement.

7.5 NETWORKING INFRASTRUCTURE

Up to this point, the focus has been on selecting the best infrastructure services and components to meet the needs of the solution. Now the focus shifts to how the components connect and communicate with each other to carry out the activities of the solution.

Throughout the process of solution technology definition there is a degree of iteration required. The level and frequency of iteration may vary from step to step, but the aim of achieving the functional requirements and NFRs is paramount. Iteration occurs each time an infrastructure service, provided by an infrastructure component, is proposed as part of a solution. This triggers a review of the solution requirements to ensure that they are all still being met. This is a form of static regression testing.

This approach continues throughout solution technology definition, but the activity of defining the networking infrastructure is a boundary or watershed in the process. At this point any trade-offs between infrastructure services and components should have been resolved so that the focus can move on to connectivity and interfaces.

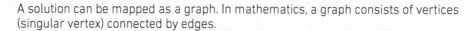

Solution as a graph

A solution can be mapped as a graph. In mathematics, a graph consists of vertices (singular vertex) connected by edges.

For a solution, each component or SBB is a vertex, and each interface is an edge. The number of edges may be counted on the graph or may be calculated using the degree sum formula.

The degree of a vertex is the number of edges it is connected to. If these are added up for all vertices and then divided by two, the result is the number of edges. Thus, if the number of interfaces is recorded for each SBB, then the total number of interfaces can easily be calculated.

In the graph shown in Figure 7.3, there are:

- 13 SBBs in total
- 6 with index 1 = 6
- 3 with index 2 = 6
- 2 with index 3 = 6
- 1 with index 4 = 4
- 1 with index 6 = 6
- 28 total indexes
- 14 total SBBIs

Figure 7.3 Solution as a graph

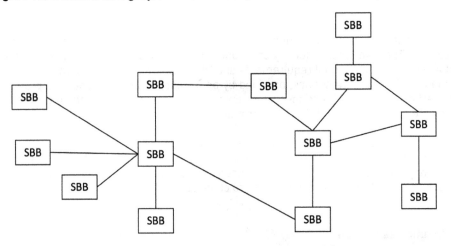

Each interface needs to be examined individually to determine the correct infrastructure service requirement. Interfaces are about communication, so the requirements will concern the type and volume of the communication traffic that passes across the interface from one SBB to another. Note that not all interfaces require networking – for example, an interface between two processes may be entirely manual. However, in the interest of completeness, each interface should be considered even if it can be swiftly ruled out.

Note that interfaces between SBBs are known as SBBIs.

The process here is similar to mapping SBBs to infrastructure services. First, a list of SBBIs that require the use of infrastructure services is prepared. The services are mainly in the networking and communications category. The individual requirements for each interface are then identified and clarified. These will largely be NFRs but there may be some functional requirements about the message contents, for example. This information allows suitable infrastructure services to be selected to support each interface.

A RACI matrix (see Table 7.4), may be used to show stakeholder involvement in this activity.

Table 7.4 RACI matrix: Defining the networking infrastructure

Activity	Infrastructure architecture	Solution architecture	Business sponsor	Enterprise architecture
Itemise and categorise inter-SBB interfaces (SBBIs)	RA	CI		
Identify infrastructure service requirements	RA	CI		
Categorise infrastructure services as existing, requiring modification, new or replacement	RA	CI		
Validate that all business requirements are fulfilled by the identified infrastructure services	CI	RA		C
Authorise capital and recurring costs for the provision of new or modified infrastructure services	CI	CI	RA	I
Clarify enterprise directives relating to providing new or modified networking services	RA	I		C

7.6 FUNCTIONAL REQUIREMENTS AND NFRS

At every stage in the process of solution technology definition, the business requirements have been checked and validated by the solution architecture team to ensure they have not been adversely impacted by any of the decisions about the use of infrastructure services. Business requirements represent the quality of the solution and are the test against which it will be judged by its stakeholders.

This results in a set of infrastructure components connected by networks that can deliver the requirements of the solution.

However, the nature of architecture is that everything is connected, meaning that a design decision in one area of the solution affects everything else. For example, the decision to select a type of data store, such as a NoSQL graph database, has benefits for analytics that may support a functional requirement but is not so good for one of the NFRs such as data integrity. In other words, design decisions involve compromises that can involve trading off between requirements.

The aim of this activity is to test the design to identify any weaknesses in support for specific functional requirements or NFRs, and then modify the design to try to strengthen support, in a balanced way, for all the requirements.

It is important to recognise that testing at this stage does not mean putting software or hardware through its paces or setting up test environments, nor does it involve recruiting end users and business actors to try out the solution. The type of testing used here is called static testing and is based purely around the solution design and its supporting documentation.

Some static techniques that can be used here are:

- **Traceability:** to ensure that all requirements are present within the design.
- **Business scenarios:** realistic end-to-end use cases of the solution by business actors that illustrate the problem being addressed by the solution.
- **Data and information flow analysis:** close examination of the movement of data and information to ensure it meets or supports the requirements.
- **Control flow analysis:** inspection of decision points in the solution to ensure sufficient accurate information is available to make the correct decision, and that decision points cannot be bypassed in a way that would compromise the requirements.

The majority of testing is conducted by the solution architecture team with support from the infrastructure architecture function. However, there may be benefits in involving other stakeholders. For example, a representative of the business, such as a **business analyst**, can give another perspective on what can otherwise become quite a technical activity. Another useful perspective may be introduced by the use of **professional testers** who are familiar with the techniques of static testing and can help with creating scenarios, tracing requirements and ensuring a high degree of test coverage.

The **requirements catalogue** is the reference point for all types of requirements and should be used as the source of information during this activity.

A **requirements traceability matrix** acts as an intermediate artefact to link requirements to solution components – that is, the SBBs and SBBIs.

Existing **viewpoints** can be used to produce views based on the proposed infrastructure services and components in the solution technology definition.

A RACI matrix, such as the example in Table 7.5, may be used to show stakeholder involvement in this activity.

Table 7.5 RACI matrix: Refining solution design to achieve functional requirements and NFRs

Activity	Infrastructure architecture	Solution architecture	Business analysis	Professional testing	Other stakeholders
Test the solution design to ensure functional requirements are achieved and revise the design if necessary	I	RA	C	C	
Test the solution design to ensure that all NFRs can be achieved using the proposed solution technology definition	CI	RA	C	C	
Itemise and document tolerances for each NFR	RA	CI			
Repeat previous solution technology definition activities to achieve a balanced design that meets all requirements	R	RA			C
Ensure all business requirements are met		A			

Fallowdale Hospital patient communications: refining solution design to achieve functional requirements and NFRs

The patient communications solution contains an SBB called the Communication Management System (CMS) that requires information about patient addresses, clinic bookings and patient preferences. These are stored in a data centre using two different data storage services: a relational database management system (RDBMS) and a 'not only SQL' (NoSQL) system.

The data centre is connected to the hospital by a high-speed data connection (10 GB/s) and the CMS is connected to the hospital LAN (1 GB/s) (see Figure 7.4).

The CMS makes multiple calls to the data storage services. First it queries the clinic bookings to see any new bookings where no invitation has been sent. Then for each booking it obtains the patient preferences for the patient or named contact (parent, guardian, or carer) if appropriate. Finally, the contact details for each patient are retrieved.

This approach works in theory as the data connections are fast, but it has been noted that the use of the high-speed data connection varies during the day and there are particular points where it is used for uploading 3D imaging data, which can lead to a drop in availability.

This has led to some concerns about meeting the NFRs for data integrity and currency if there are multiple requests for information.

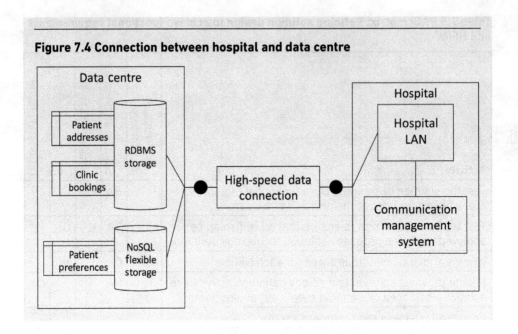

Figure 7.4 Connection between hospital and data centre

Activity 7.4

Given the information about the Fallowdale Hospital patient communications solution provided above, suggest two options for strengthening the design of the solution to lower the risk of the high-speed data connection compromising the NFRs for data integrity and currency being achieved.

7.7 END-TO-END SECURITY

Information security is an important aspect of solution design and applies in almost every line of business, but particularly so where confidential information relating to customers or commercial activity is concerned. Cybersecurity is also a big concern and this must be reflected in solution design by including robust protection against attack by malevolent parties.

Cybersecurity focuses on protecting computer systems from unauthorised access or being otherwise damaged or made inaccessible.

Information security is a broader category that looks to protect all information assets, whether in hard copy or digital form.

In a solution, there are a number of ways that security can be specified:

- **Functional requirements:** security controls that are part of the functionality may be specified as requirements and may be implemented using techniques such as role-based access control (RBAC), business rules that may be configured to prevent unwanted activity or security auditing.

- **NFRs:** measures that prevent malicious attacks and attempts to undermine the aims of the solution.

- **Enterprise directives:** principles and policies that define how security measures should be applied in specific areas of the business.

- **Technical standards:** specifications that have been adopted by the business to guarantee types and levels of behaviour that underpin security and are widely recognised as being of value externally by, for example, customers and partner organisations.

Functional requirements related to security should by now have been refined as part of previous activities to a point where they are acceptable to all stakeholders. However, there are some cases where there is some overlap with security NFRs and it is as well to include all aspects of security in the end-to-end assurance step.

In previous steps of solution technology definition, other stakeholders have been consulted and provided input, either by virtue of their accountability for the outcome or because of their specialist knowledge about the operational domain. Security aspects of a solution require highly specialised and current knowledge because of the changing nature of the threats faced. Recognising that security assurance is a complex and specialised field, organisations are increasingly convening a **security design authority** to make decisions about the security aspects of all solutions within the organisation or enterprise. Much of the accountability for assuring end-to-end security is delegated to this group.

A RACI matrix (see Table 7.6) may be used to show stakeholder involvement in this activity, including the security design authority, if there is one.

The security design authority performs two tests on a solution and provides an assessment that is acted upon by the solution architecture team and the infrastructure architecture function:

- **Compliance with directives:** auditing the solution design against security principles, policies and standards that have been adopted by the organisation or enterprise, such as ISO 27001 (ISO/IEC 27001:2013, 2013).

- **Vulnerability testing:** modelling threats and malicious activity to test the security resilience of the solution using libraries of identified threats such as Common Weakness Enumeration (CWE) and Open Web Application Security Project (OWASP) and frameworks such as Control Objectives for Information and Related Technologies (COBIT) and National Institute of Standards and Technology (NIST).

Table 7.6 RACI matrix: Assuring end-to-end security

Activity	Infrastructure architecture	Solution architecture	Security design authority	Enterprise architecture	Security architecture and specialists	Business sponsor
Constitute the security design authority and ensure effective decision making	C	C	I	RA	C	I
Present the solution design and architecture description with all relevant views, business scenarios and other artefacts	I	RA	I			
Present proposed changes to the infrastructure architecture and the proposed solution technology definition	RA	I	I			
Assure the solution complies with all security directives	I	I	RA	C		I
Test all aspects of the solution for security vulnerabilities	I	I	RA	C		
Obtain evidence and detail about specialist security areas	I	I	RA	C	C	
Provide a security risk assessment with any recommended security measures or actions	I	I	RA	C	C	I
Redesign solution to strengthen security	R	RA	CI	CI	C	
Modify solution technology definition	RA	R	I	CI	I	
Confirm assurance of end-to-end security	I	I	RA	I		I

Fallowdale Hospital patient communications: assuring end-to-end security

Part of the proposed patient communication solution involves collecting communication preferences directly from the patient using a website. One security aspect of this process is that credentials are provided by the patient and are then matched to patient demographic data in order that the patient can be identified, and their preferences associated with the correct records (see Figure 7.5).

Figure 7.5 Updating patient communication preferences

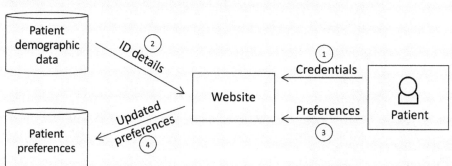

The security design authority performed an assessment of this part of the solution design to check that it complied with all relevant enterprise security directives. The findings were as follows:

- Patient demographic data is classified as personally identifiable information (PII) and therefore has a highly confidential rating in the information security framework.

- User Access Provisioning policy states that access to any asset (information or system) must follow a defined process in accordance with ISO 27001 (ISO/IEC 27001:2013, 2013).

- Management of Secret Authentication Information policy states that this must be controlled through a formal management process in accordance with ISO 27001 (ISO/IEC 27001:2013, 2013).

- Identity and Access Management (IAM) policy states that approved information security controls must be used to establish the identity of end users and business actors and to define their level of access in accordance with ISO 27002 (ISO 27002:2013, 2013).

- Information Access policy states that all access to data must be logged and available for audit.

- A suitable IAM solution has been identified in the infrastructure technology catalogue and TRM that has been approved for use with end users and business actors who are external to the business.

The security design authority recommended the solution design be revised to take into account these findings. They also noted the following points that would appear in a vulnerability report:

- Access control should follow the principle of least access (granting access to the minimum set of resources with the lowest privilege level to each actor that allows business activities to be performed).

- Solution is vulnerable to an injection vulnerability due to passing information provided by a user to a database (OWASP, 2021: A1; MITRE Corporation, 2021).
- Solution has insufficient logging and monitoring that is identified as a vulnerability (OWASP, 2021: A10; MITRE Corporation, 2021).

Activity 7.5

Based on the security design authority's assessment of the design for updating patient communication preferences, make a list of recommendations for redesigning the solution.

7.8 GAPS IN INFRASTRUCTURE SERVICE PROVISION

A key output of the solution technology definition process is a report on any **gaps** in the provision of infrastructure services that would **hinder or prevent** the **implementation** and **deployment** of the solution.

Key decisions about the implementation of a solution cannot be taken without this information. For example, service gaps are likely to add to the costs of implementing and deploying the solution, and costs are a major factor in the decision to proceed with any project or programme.

Costs may be accrued by a solution in multiple ways, including those costs associated with solution architecture and design, implementation, deployment and ongoing management and maintenance. The marginal costs linked to infrastructure may be isolated by itemising the gaps in infrastructure service provision and analysing the associated costs.

Exposing a higher-than-expected level of costs may lead to a rethink of:

- the infrastructure technology used to support the solution;
- the architecture and design of the solution;
- the scope and scale of the solution;
- the inclusion of baseline opportunities.

Whether changes are made in any of these areas or not, stakeholders will be given the opportunity to validate the solution technology definition and may insist on a revision.

The other benefit of identifying and itemising gaps in infrastructure service provision is that the resulting requirements for change must be part of the solution delivery roadmap.

The process for achieving the list of changes and costs is similar to the gap analysis that was performed between the baseline and target solution architectures during the validation phase of the solution architecture life cycle. Here the list is constructed by comparing the solution technology definition and the infrastructure technology catalogue.

Each item may be categorised as follows:

- **New:** service is not currently provided.
- **Replacement:** existing service replaced by an alternative.
- **Modification required:** service exists but needs reconfiguration, rescaling or expansion into a new geographical region, for example.
- **Removal:** service is no longer required.

Each of these changes needs to be assessed to establish the size and scale of the change and what will be required in terms of cost, resource and time. Note that although all changes require effort and this is likely to involve some cost, some changes may reduce ongoing costs. For example, in the case of the replacement of one service by another, there will be some cost in making the switch, but the running costs of the new service may be less due to improved efficiency. The entire basis and rationale of some solutions is to reduce ongoing costs, but this can rarely be achieved without investing in change first.

7.9 VALIDATING THE SOLUTION TECHNOLOGY DEFINITION

Once all the steps of the solution technology definition have been completed, all the necessary information is ready to be presented to the stakeholders for validation.

Many stakeholders will have been directly involved in the process and others will have been consulted during decision making or informed of the outcome of those decisions. If stakeholder communication is effective, there should not be too many surprises during the validation step. However, it is an opportunity for all stakeholders to focus on the infrastructure and technological aspects of the solution and see all the relevant information in its completed form.

This step is jointly led by the solution architecture team and the infrastructure function. The information may be delivered to stakeholders electronically for review and followed up with a workshop at which there is an opportunity for discussion and questions.

A solution technology definition validation workshop typically covers:

- A walkthrough of the solution technology definition with all stakeholders present or represented.
- A description of all changes and revisions of the solution architecture.
- A breakdown and explanation of the infrastructure costs and any savings associated with the solution.
- A discussion of the tradeoffs behind the infrastructure choices.
- Raising and addressing any concerns, either during the workshop or following further investigation.

REVIEW QUESTIONS

1. Which step is performed first in solution technology definition?

    a. Understand baseline opportunities.

    b. Clarify precursors and requirements.

    c. Assure end-to-end solution security.

    d. Map building blocks to infrastructure services.

2. How is an infrastructure catalogue used in solution technology definition?

    a. As a source of infrastructure services for a solution.

    b. As a source of requirements for a solution.

    c. To define the interfaces between solution components.

    d. To define the end-to-end security for a solution.

3. A solution architecture team has completed the logical design and validation phases of the solution architecture life cycle. The solution is now being put through the solution technology definition process. What **two** possible events during this process will require the solution design to be revised?

    i.   A security vulnerability is identified.

    ii. A solution component is mapped to an infrastructure service.

    iii. A target architecture constraint is breached.

    iv. An infrastructure service capacity gap is identified.

    a. i and ii only.

    b. iii and iv only.

    c. i and iii only.

    d. ii and iv only.

4. Which type of solution component is **most** likely to map to a networking infrastructure service?

    a. Non-functional requirement (NFR).

    b. Solution building block (SBB).

    c. Architecture building block (ABB).

    d. Solution building block interface (SBBI).

8 IMPLEMENTATION

LEARNING OUTCOMES

When you have completed this chapter, you should be able to demonstrate an understanding of the following:

- Building the delivery roadmap
- Monitoring and supporting delivery of the solution
- Validating product delivery
- Agreeing service levels, targets and metrics
- Supporting future change and maintenance
- Communication with the business

8.1 BUILDING THE DELIVERY ROADMAP

The roadmap development phase is a watershed in the solution architecture life cycle. It is where the carefully architected and validated solution design is handed over to be implemented and deployed by a variety of specialist teams. These delivery teams will be managed by P3M, and the delivery roadmap is their contract with the business and its stakeholders.

After the handover, the role of solution architecture changes significantly. Up to this point the solution has been in the hands of the solution architecture team and the main focus has been on logical design. From this point onwards, there is little or no requirement for this type of work, and solution architecture takes on a governance role, becoming the point of reference if anything relating to the architecture or design is unclear, or needs to be changed.

The roadmap is a critical document that serves to communicate at a high level and to a broad audience what needs to be delivered and in what sequence:

- **Stakeholders:** can focus on business benefits and ensuring the correct sequence and time frame for their achievement.

- **P3M:** can see the blocks of work that need to be broken down into programmes and projects to deliver the benefits.

- **Delivery teams:** can understand the context in which their implementation work is to be completed and integrated.

- **Solution architecture team:** is able to validate that the SBBs will be created and integrated in the correct sequence to deliver the business benefits.

8.1.1 Roadmap structure

Delivery roadmaps are needed to portray complex, multi-dependent change over a medium to long time period. The roadmap therefore shows only the most important **deliverables** that are easily recognised by all stakeholders, and places them within **broad time periods** – quarter years or months, for example – that are widely in use within the organisation for activities such as resource planning and budgeting. These are shown as vertical divisions. Delivery milestones and other significant events may also be shown (see Figure 8.1).

Deliverables may also be organised into **workstreams** that are normally shown at right angles as horizontal divisions and help to make the roadmap easier to understand. They may be based on a number of factors, such as the business area or geography of the change activity represented.

Deliverables are planned and organised by P3M using a combination of **programmes** and **projects** (see Figure 8.2). A programme is a collection of related projects and may be linked directly to a roadmap workstream. Some deliverables are of a sufficient size and degree of complexity to require a programme. More straightforward ones are usually delivered as part of a single project.

It is interesting to note that **SBBs** may relate to either **deliverables** or project **work items**. This depends on the granularity of the SBB. Some are small, fine-grained components whereas some are major parts of the whole solution, sometimes known as coarse-grained components.

Fine-grained SBBs appear as work items in project plans and are not visible on the roadmap. Coarse-grained SBBs that are made up of many fine-grained components working together are more likely to appear on the roadmap as deliverables.

To appear on the roadmap, an SBB must be separately deliverable, recognisable to stakeholders, and be associated with a business benefit or capability.

Figure 8.1 Roadmap structure

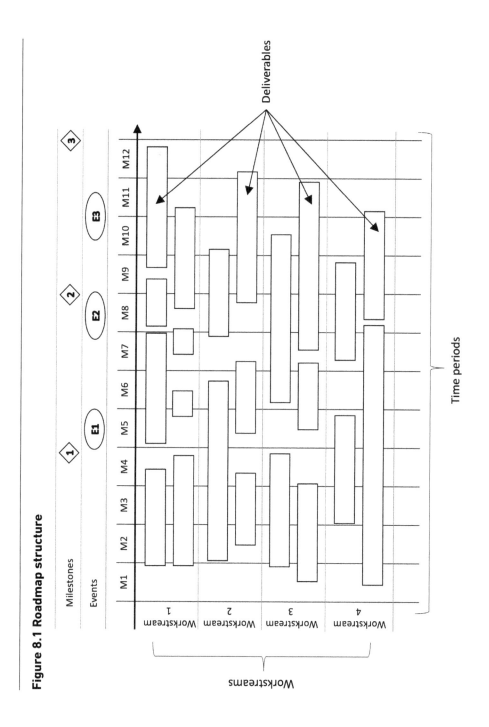

Figure 8.2 Delivery roadmap metamodel

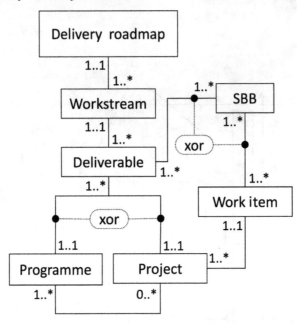

Fallowdale Hospital patient communications: fine and coarse-grained components

Three SBBs that are present in the patient communications solution architecture are:

- **Preference data:** considered fine-grained and not visible directly on the roadmap, although it is a critical component of **preference management** and is also part of the **patient website** and is one of the items collected during **patient data acquisition**.

- **Message content management system:** is coarse-grained, appearing on the roadmap as a deliverable, and is made up of several finer-grained components including **general messages, response details, business rules** and **attendance instructions**.

- **Waiting list processes:** also coarse-grained and appears on the roadmap. Its finer-grained components are shown in diagram form in Figure 8.3.

Figure 8.3 Coarse and fine-grained component example

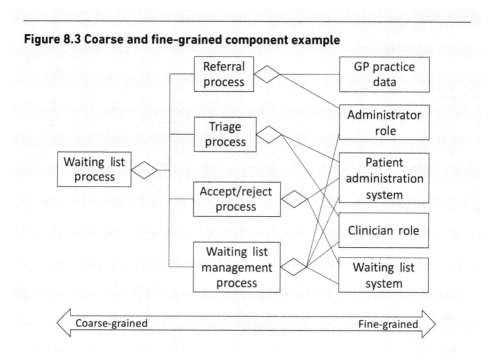

8.1.2 Inputs to the roadmap

The delivery roadmap represents a transition from the logical realm of architecture and design and is where the solution begins to be transformed into a physical reality. The shift from logical to physical is also evident in the **solution technology definition** activity where SBBs and SBBIs are matched up with technology components from the infrastructure architecture domain.

Significant technology components that need to be acquired or modified may appear on the roadmap as deliverables. Others that are components of a deliverable can be found by drilling down into project plans.

Behind the roadmap, P3M develops detailed plans with more specific deliverables, dates and dependencies, which align closely with the roadmap. Additional date tolerances are often included in the roadmap to allow for the compound dependency risks of multiple components needing to be delivered to support a single roadmap item.

The main inputs to the roadmap are:

- **Target architecture:** solution design documents that show how everything works together.

- **Gap report:** provides an itemised list of SBBs and SBBIs where change is required.

- **Solution technology definition:** provides an itemised list of infrastructure components where change is required.

- **Existing dependencies:** planned activity in other programmes and projects that overlaps with the solution and could impact its delivery.

8.1.3 Development process

The solution architecture team, working with P3M, are responsible for the process of building the delivery roadmap. Solution architecture provides three of the four inputs (target architecture, gap report and solution technology definition), with P3M providing the fourth (existing dependencies).

The development steps are:

- collect and validate inputs;

- identify and agree deliverables;

- perform outline planning to manage dependencies and resources;

- trade off between stakeholder expectations;

- produce draft roadmap;

- present to the business (possibly via the design authority);

- refine and circulate for approval.

Once a draft roadmap has been developed, this should be discussed with the business sponsor to ensure it is compatible with the priorities of the business.

When P3M, solution architecture and the business sponsor are happy with the roadmap, it can be presented to a broader group representing all the stakeholders for the solution. At this point, stakeholders may raise concerns about the dates or sequencing of deliverables, and this can lead to adjustments being made. Any modifications must be done with the agreement of the business sponsor.

8.1.4 Turning SBBs into roadmap deliverables

The major constituents of the roadmap are the deliverables, time periods and workstreams. The most important of these to identify are the deliverables and this should be done at the start of the process. Deliverables are derived directly from SBBs and are formed by grouping them together logically to deliver specific business benefits. Note that a single SBB may be part of multiple deliverables but only needs to be developed or procured once, although it may need configuration.

This approach of grouping together SBBs into deliverables may be performed using several techniques, such as affinity mapping, but a popular method is to produce a type of tree diagram called a **work breakdown structure (WBS)**. This technique is widely used in project planning and begins with the highest-level product or the end result that the project is trying to achieve. For developing a solution delivery roadmap, the highest-level product is the solution itself.

Fallowdale Hospital patient communications: identifying deliverables for the roadmap

The solution architecture team has worked with P3M to produce the WBS shown in Figure 8.4.

There are four top-level deliverables that are required to communicate effectively:

- **Address:** where to send the message.
- **Consent:** agreement to receive communications and preferred methods.
- **Message contents:** what will be sent.
- **Message delivery mechanism:** how the message will be sent.

Figure 8.4 Patient communication WBS

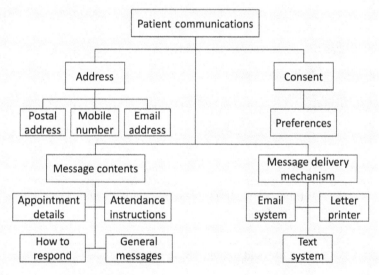

The method is then to identify the next level of item that is required to deliver the solution. Further levels can be added until all solution components have been included.

Note that this technique may also be used much earlier in the solution architecture life cycle as a way to analyse a problem and start thinking about what might be required to solve it.

A WBS generally shows top-level deliverables and one or two levels below that. The diagram would become hard to follow and to maintain if much more detail were added.

Its value is as a guide to identifying deliverables that need to be broken down further and organised into programmes and projects.

Fallowdale Hospital patient communications: identifying lower-level delivery components in the WBS

On the patient communications solution WBS, there is a component deliverable below 'Message contents' called 'General messages'. These are messages that need to be included with every communication. At present these are added manually to letter templates, but this is done inconsistently, with the result they are often missed out or sent inappropriately. Following discussions with a data architect and a business analyst, it is clear that a system needs to be designed and built to store and retrieve general messages and that procedures need to be designed to update them. It will also be necessary to import existing general messages into the storage and check how they are to be used with the business.

This analysis identifies four delivery components which comprise 'General messages':

- Storage and retrieval of general messages.
- Procedures to update general messages.
- Importing existing messages.
- Business rules for general messages.

Activity 8.1

The Fallowdale Hospital patient communications WBS shows 'Consent' as a top-level deliverable for the patient communications solution, with 'Preferences' below it as a component deliverable.

What components and supporting activities are needed to enable patient communication preferences to be used in the new solution?

Are there any dependencies between them that would affect the order in which they are delivered?

8.1.5 Project and programme planning

Having confirmed what needs to be delivered, the P3M function can start to think about how this could be achieved. The overarching responsibility of all those involved is to deliver the solution in line with the needs of the business, which is represented by the stakeholders. Therefore, the priorities and concerns of the stakeholders should be the main drivers to determine the sequence of delivery.

The stakeholders, led by the business sponsor, will have the opportunity to review, amend and approve the roadmap before the implementation begins. The solution architecture team, having worked closely with stakeholders up to this point, should have a very clear idea of their priorities.

There are many factors involved in the development of project and programme plans apart from the wishes of stakeholders, including:

- **Estimation:** how quickly tasks can be completed.

- **Dependency:** the natural or logical order in which tasks must or should be completed.

- **Resource allocation:** how resources can best be used to deliver the desired outcomes.

- **Risks:** how to minimise or mitigate all types of risk during the project.

The solution architecture team can advise on the business risks that were identified during the business case option selection process. Other information from the business case may also be relevant, such as the return on investment analysis. Stakeholders can also advise on general risks in other areas of the business.

Risk management is the responsibility of the P3M function throughout the implementation and deployment of the solution and a formal process will be used. During the roadmap development process, only those risks that are relevant to the sequencing of delivery are taken into account, although all identified risks should be captured for future reference.

The P3M function can then look in more detail at the programme and project plans to see if there are any serious concerns with the roadmap. Delivery team representatives may be consulted during this review. Any issues should be raised with the solution architecture team and can be escalated to the business sponsor.

Another point to agree upon is the schedule of communications and updates that are part of the governance of the delivery of the solution. This schedule should be published, along with the roadmap, to all solution stakeholders.

Fallowdale Hospital patient communications: delivery roadmap

The delivery roadmap for the Fallowdale Hospital patient communications solution is shown in Figure 8.5.

Note that the roadmap has two workstreams for business and technology components and an additional summary workstream showing the capabilities that will be achieved. Three milestones have been scheduled at roughly equal intervals when a major review of progress will take place. The first is shortly before a board meeting at which an executive summary has been requested. The second is timed to occur before the end of the financial year. It is hoped that the solution will be highlighted in the hospital's annual report.

Figure 8.5 Patient communications solution roadmap

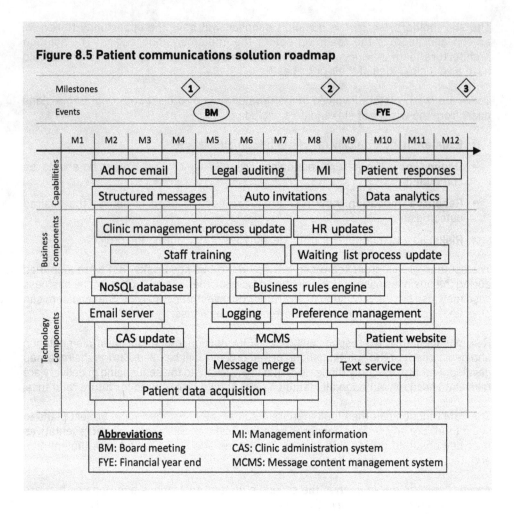

8.2 MONITORING AND SUPPORTING THE DELIVERY

There are a number of levels of involvement for the solution architecture team once the work of implementation has commenced.

At one extreme, solution architecture has little involvement beyond attending regular review meetings. Responsibility for ensuring that the solution adheres to the agreed architecture and design is largely delegated to P3M and any divergence is raised directly with the business to decide how to proceed.

A more formal approach to governance places solution architecture in the role of guardian of the design that represents the ideal for the business, having been fought out between the competing interests of multiple stakeholders. The benefit of a greater role for solution architecture in the governance of delivery is that the implemented solution is closer to the design and, where it varies, it is closer to the requirements.

In general, more complex designs and those with greater business impact should be dealt with more formally. Simpler, smaller-scale solutions can largely be delegated to P3M.

8.2.1 Role of the design authority

Throughout the work of architecting and designing the solution, the design authority has provided a focal point for decision making and stakeholder communication. The design authority:

- is fully representative of stakeholders;
- has access to specialist knowledge when required;
- provides updates to all or selected stakeholders;
- receives queries and concerns which can then be dealt with directly or escalated.

During the completion phase, the design authority is ideally placed to continue these activities. The differences between the completion phase and the previous phases of the solution architecture life cycle are chiefly ones of scale and complexity. Up to and including the roadmap phase, the focus has been on a single design being produced by a single team, albeit with the involvement of multiple stakeholders (see Figure 8.6).

The major organisational change during the completion phase is that the P3M function takes over coordination of the delivery of the roadmap elements, having been authorised by the business sponsor. As the delivery programmes and projects progress, P3M can keep all stakeholders fully informed by reporting progress to the design authority.

Figure 8.6 Role of design authority during the completion phase

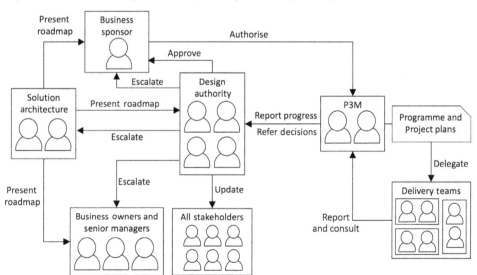

Any design decisions that exceed the specifications given to a delivery team by P3M can be referred to the design authority, which can escalate decisions if necessary:

- To **solution architecture** if changes to the architecture and design of the solution would be needed.
- To the **business sponsor** if business requirements or constraints would be affected.
- To **business owners and senior managers** if business operations are likely to be impacted.

Decisions that have been approved by the design authority need to be authorised by the business sponsor before P3M can amend programme and project plans and issue new specifications to the delivery teams affected by the change.

8.2.2 Complexity of design during implementation

Design activities in the completion phase are more complex, because they involve multiple components being implemented by different delivery teams with different skill sets. In addition to meeting their own specifications, they must work with other components using strictly defined interfaces. Any failure to comply puts the goals of the overall solution at risk.

There is also a difference of scale. The logical design activities of solution architecture can be completed in a relatively short period. However, there are many more constraints on physical activities, such as the development of software, acquisition of technology, or redesign of processes and organisational units. In general, this means delivery timescales are longer and resource use greater.

The number of dependencies increases exponentially with the number of components. This effect can be limited by packaging as many components together into a single deliverable so that the dependencies can be dealt with internally by the delivery team.

The solution architecture approach aims to reduce complexity so that the solution can be delivered with the maximum possible benefit to the business and the minimum disruption to operations. This is achieved through:

- analysis of requirements;
- design of independent components;
- clear specification of interfaces;
- reuse of existing components.

8.2.3 Scheduled meetings and regular reporting

Since there are multiple components being developed that may require decisions to be referred to the design authority, it is usually best to set dates when the members meet to consider these and any other relevant matters. This approach makes good use of the time and resources of the design authority members and any expert witnesses who are required to attend.

A schedule of regular meetings helps the development and delivery teams to coordinate their own activities because they know when critical decisions will be made. Therefore, planning sessions, supplier demonstrations, user testing and other activities that may throw up questions that need to be referred to the design authority can be scheduled before a meeting date and the work that would result from a decision scheduled afterwards.

An important way to monitor progress in the implementation and delivery of a solution is the use of regular reports from the teams responsible for developing or acquiring solution components.

This is a standard approach within P3M, where regular reviews of progress are held at programme and project meetings. These often have an internal and external aspect where minor technical aspects are dealt with internally by the P3M community before a presentation of progress is made to the business.

The solution architecture team should be represented at these progress updates, because there are often discussions about prioritising activities or allocating resources that can have an impact on the roadmap. Any such changes should be referred to the design authority or the business sponsor for a decision and approval.

Any question of changing the scope of a deliverable so that a specification or requirement will not be delivered should be referred to the design authority before any action is taken.

Some members of the design authority may be present at P3M-led progress update meetings. Other members can be kept abreast by receiving update summary reports. It is quite usual to have a report on progress as a standing agenda item at design authority meetings, at which point any queries can be raised and dealt with. Queries that cannot be dealt with from the summary information provided may be referred back to P3M.

8.2.4 Altering the roadmap

The summary report from P3M may highlight risks to the delivery of one or more roadmap items. If so, the report should be accompanied by details of the problem and the risk management activities that are being used to avoid or mitigate the risk.

In cases where the risk is unacceptably high, it may be necessary for the roadmap to be altered. This is a very serious matter as the delivery roadmap is a contract with the business stakeholders, who may have based operational activities and strategic plans around it. Before the roadmap can be altered, therefore, the business stakeholders must be fully involved. It is critical to begin the process of rebuilding the roadmap as soon as it has been established that the risk is significant. This is true even if it just involves a short delay with no knock-on effect on other deliverables.

Fallowdale Hospital patient communications: roadmap alteration

The following notifications have been received from P3M and are being put on the agenda for discussion at the forthcoming design authority meeting:

- The logging system will be delivered earlier than expected. Both the legal auditing and MI capabilities are enabled by the logging system.

- There is a delay to the delivery of the message content management system (MCMS) because of some technical issues. The current estimate for delivery is 3–6 weeks late, moving the end point to month 10 (M10) on the roadmap. A manual work-around has been identified that would mean the use of back-office IT staff to update message contents.

Activity 8.2

Consider the notifications received from P3M about the logging system and the MCMS. For each notification:

- Identify the implications for other roadmap items.

- Describe some possible decision options.

- Do any of these decisions need to be escalated and if so, to whom?

8.2.5 Vetting designs

The delivery of the solution is managed by P3M in terms of allocating the work to a diverse number of delivery teams and monitoring their progress. P3M may have some expertise in the work being done, for example, being able to understand sprint plans and test reports produced by software development teams. However, when it comes to architecture and design, it is usual to refer any assessment or vetting to a body such as the design authority or another technical committee. Such a body can organise for the necessary expertise to be made available.

Design assessments may be made by architects and specialists from the appropriate domain, such as security, business, or software, and may work together to check that the design meets the needs of the solution:

- **Effectiveness:** meets all requirements and conforms to standards and constraints.

- **Efficiency:** makes best use of resources.

- **Encapsulation:** only uses defined interfaces and follows their specifications.

8.3 VALIDATING PRODUCT DELIVERY

The delivery of a product – meaning an SBB, component, or other deliverable – indicates that it is completely ready for deployment. Validating product delivery means establishing it is of sufficient quality and meets its specification and acceptance criteria. As more products are delivered, the interfaces between them can also be validated to ensure that they will work together when deployed.

Delivery teams report successful delivery to P3M, who then report back to the business via the design authority or a programme board. Apart from confirming that delivery dates accord with the roadmap, stakeholders need to have confidence in the quality of the products being delivered.

Governance of product quality has a number of strands that fall into different areas of responsibility:

- **P3M:** primarily concerned with time, cost, resources and to a certain extent the scope.

- **QA:** aims to ensure robust processes are in place that ensure a high-quality product is delivered.

- **Testing:** closely examines products on delivery and during development, making a formal assessment of quality.

- **Business change management:** concerned with how easily the product can be deployed within the business and the deployment of all business components of the solution.

There is some overlap between these strands, with testing as the common factor. For example, in projects involving software development, the project manager will often insist on a development life cycle being followed and its activities integrated into the project plan. A life cycle is also a control measure used by QA professionals as it contains processes that promote good practice with regard to quality. Testing is a major part of software development life cycles. Ease of deployment includes usability that is often expressed as an NFR and therefore subject to testing.

A concern for solution architecture is that the governance of product delivery has a disproportionate focus on software development. This is especially true of testing, which is either much more limited or non-existent for non-software components of a solution. Solution architecture provides adequate specifications that would allow the same standard of testing as for software:

- **Externally sourced:** components obtained from equipment suppliers, off-the-shelf software and even externally developed software is often subject to much lower standards of governance and testing. Solution architecture must ensure these are tested to ensure they meet the acceptance criteria by professional testers, as listed above. This needs to be emphasised to P3M so that it is included in project and programme plans.

- **People and organisation units:** where roles and responsibilities are changing, plans may include various forms of communication, including training, but its effectiveness is rarely tested. This comes under business change management, as described above, but testing may involve professional testers working with HR and other business managers.

- **Processes:** quite often lumped together into training courses and other forms of communication, the actual operation of processes often goes untested despite the many functional requirements and NFRs that apply to them. This can be done using scenario-based testing that is frequently used to test software systems.

- **Information:** sometimes tested in conjunction with any software that uses or manages it, but many of the requirements and constraints that are put in place to ensure immediate and long-term quality are omitted from testing programmes. Solution architecture needs to ensure that all information and data components within the solution have sufficient test coverage.

A factor that can impede the correct validation of product delivery is change management and version control of the specifications that deliverables are being built to meet. It is possible for a specification to change during implementation.

Any change should be subject to strict governance with decisions being made and authorised at the appropriate level. Failure to follow change control procedures could result in one version of a specification being used to design and build a product and a different specification being used for testing and validation.

Fallowdale Hospital patient communications: delivery validation

The **reschedule appointment** process has been redesigned and documented and is completely ready for deployment, which will take place through staff training and the provision of reference material.

The following details of its acceptance criteria have been provided as part of a validation audit by the project manager and made available to the design authority:

- **Trigger event:** request from patient.
- **Included processes:** cancel appointment, book appointment.
- **Input:** identified patient, date, time and clinic details for current appointment, available future appointment dates.
- **Output:** new empty slot, new booking or new waiting list entry.
- **Interfaces:** clinic management system (CMS), clinic administration staff.
- **Testing:** all scenarios tested (available on request), messages sent using correct preferences, clinic schedules correctly updated, waiting lists correctly updated.

8.4 AGREEING SERVICE LEVELS, TARGETS AND METRICS

An SLA is a commitment between two parties: the service provider and the client (sometimes called the service user). SLAs define the normal level of provision and what will be done if this is not maintained. The service provider is usually an external organisation such as a supplier, for example an internet service provider (ISP), or can be an internal business unit such as the IT department. Similarly, the client can be internal or external to the business.

Cloud provider SLAs

Where IT services and RPA are provided by a cloud provider, SLAs are used as the basis for billing within defined limits. This means that when provision varies by volume or quality due to demand or time schedule, the client agrees to pay the supplier accordingly.

A simple way of looking at SLAs is that they directly mirror the NFRs that were specified for the SBBs that make up the service.

One decision that needs to be taken is the scale of the service that an SLA will be applied to. This determines the metrics that will be used to monitor the service and feed any action or escalation.

Another decision is the frequency of measurement. With automation, measurements can be made at vanishingly small intervals, approaching continuous monitoring. Whether this is appropriate, however, depends on what is to be done with the information.

SLAs can be made up of exact figures such as speeds, percentages or volumes that are target values, but may also have additional figures that represent tolerances above or below the target figure, or both. This approach allows for multiple response options depending on the degree of variance from the target.

SLAs for business continuity and disaster recovery

As part of a business continuity or disaster recovery plan, functions and activities are typically classified as critical or non-critical for the operation of the business. A solution may be placed in one of these two categories and appropriate plans made for recovery.

For each component such as staff, organisational unit, system, technology, information, a plan must specify:

- **Recovery point objective (RPO)**: the state of the component that the plan aims to restore. This may be a point of time prior to the failure, commonly used with data and information, or a percentage of the previous capability, such as an organisational unit.
- **Recovery time objective (RTO)**: the maximum amount of elapsed time that must be taken to restore the component to the state specified in its RPO.

For the solution as a whole, a single measure is often used, known as **maximum tolerable period of disruption (MTPD)**.

Details of business continuity and disaster recovery SLAs are specified in the ISO 22301 standard (ISO 22301:2019, 2019).

Activity 8.3

The patient communications roadmap (Figure 8.5) shows seven capabilities:

- ad hoc email;
- legal auditing;
- MI;
- patient responses;
- structured messages;
- auto invitations;
- data analytics.

Select two of these and identify the service levels that are suitable for the business to operate effectively. Where appropriate, include any tolerances where additional actions such as escalation could occur. Indicate what is to be measured and how frequently this should be done.

The needs of the business with regard to a service may change over time. Requests for change to the service level need to be dealt with using a change management process in a similar way to requests for new or modified functionality. Any change can have architectural implications at both the solution and infrastructure levels and needs to be fully assessed before it is implemented.

The infrastructure architecture function may also wish to make changes to SLAs over time. For example, if the metrics show that a certain service is never used to the level specified in its SLA, infrastructure architecture may recommend a reduction in the level provided. This could free up resources that could be used elsewhere or reduce the cost of running the service.

8.5 SUPPORTING FUTURE CHANGE AND MAINTENANCE

One of the strengths of solution architecture is that solutions can be designed and built with the future in mind. Components are designed to be replaceable because their clearly defined interfaces allow alternatives to be plugged into an existing solution. Components and interfaces with fully documented functionality are also much easier to reuse within the current solution or elsewhere in the enterprise.

All the necessary details to make changes to an existing solution can be found in the enterprise architecture repository where the final versions of all artefacts and other documentation must be lodged at deployment. Other parts of the enterprise architecture repository are also updated at deployment, if affected by the solution, so that those responsible for changes made anywhere in the enterprise will have access to the latest current-state information.

There are a number of reasons for updating an operational solution and these are often categorised as:

- **Corrective:** something not working according to specification that needs to be fixed.
- **Adaptive:** a change in the operating environment of the business requires a change to the solution or new requirements have been identified.
- **Predictive:** a risk has been identified that can be avoided or mitigated for by changing the solution.
- **Perfective:** the solution can be improved beyond its current specification.

Changes may be as small as a modification to an existing component or may involve adding, modifying and removing multiple components. Any change to an existing solution represents a risk to the business. The size of the risk is roughly proportional to the size of the change and the importance of the solution to the business.

A senior representative of the business needs to authorise most changes; this could be the business sponsor for the solution. This person may have been given the role of solution owner after deployment. Corrective and perfective maintenance that does not change the solution requirements may not need authorisation and may be delegated to operational teams such as DevOps.

The majority of changes can safely rely on the artefacts and other documentation produced during the solution architecture life cycle. For example, the stakeholder and concerns registers will still be accurate, although this could change over an extended time period.

The architecture and design of an existing solution supports all types of change in a number of ways:

- Diagnosing problems systematically using solution models.
- Isolating components and interfaces that need to change.
- Identifying the impacts of change.

Operational measurements such as SLAs and KPIs can add to the evidence both for the need to change and to help identify which parts of the solution can be modified to bring about the change.

Activity 8.4

The solution architecture team has been asked to assess a change request from the operations director. Currently all hospital appointments take place in person at the hospital or a few other sites in the community. The hospital would like to offer remote appointments by video or phone.

What components of the patient communications solution would need to be modified to accommodate this change?

What additional components would be required?

Is there any impact on the service levels offered by the current solution?

Are there any additional impacts?

8.6 COMMUNICATING WITH THE BUSINESS AFTER DEPLOYMENT

During the completion phase of the solution architecture life cycle, the design authority maintains communication with stakeholders through regular updates to report progress and keep all parts of the business fully informed.

At some point after the last component of the solution has been deployed, the solution becomes part of the operational responsibility of the business and any communication about the solution takes place within the business.

Solutions that are the result of solution architecture should deliver exactly what the business requires and thereby enhance the standing of the discipline within the organisation. This should mean that the next time the business has a problem or wishes to take advantage of an opportunity, the solution architecture approach will at least be considered.

An exercise that can help to reinforce the benefits of solution architecture is to capture any lessons that have been learned during the solution architecture life cycle and circulate the findings to all involved. This can form the basis of improving or refining the solution architecture process, but also helps to embed solution architecture into business processes that relate to business change.

Additionally, any business issues or needs that were excluded from the scope of the solution can be itemised and potentially addressed in a future solution.

Monitoring and control of various aspects of the solution will have been agreed with the business and this includes escalation points and actions in case there is variance from the agreed SLAs.

In the longer term, it may be beneficial to schedule some analysis of the solution to see its effect on the business after it has been in use for a continuous period. This type of review can be helpful in identifying requirements that were not apparent when the solution was being designed.

Typically such reviews are scheduled six months or a year after the key components of the solution are in place.

Fallowdale Hospital patient communications: post-solution reviews

The following schedule of reviews (see Table 8.1) has been agreed to follow the successful deployment of the patient communications solution.

Table 8.1 Post-solution review schedule

Date	Event	Information presented
Monthly	Hospital board meeting	• Key measures and performance statistics • Summary of costs and return on investment • Ratio of messages sent by post, email and text
3 months after deployment	Lessons learned review	• Report from solution architecture • Report from project sponsor • Analysis of stakeholder feedback • Action plan

6 months after deployment	Review meeting	• Detailed analysis and summary of monthly reports
		• Ideas from business:
		  ▪ Add results from appointments and tests to solution
		  ▪ Hospital Communications department to manage message contents
		  ▪ Full integration with remote appointments
		  ▪ Investigate possible use of solution for clinical trials
		• How to improve digital take up

1 year after deployment	Review meeting	• Progress report • Points for annual report • Ideas for future solutions

REVIEW QUESTIONS

1. A new solution has been developed to capture and filter citizen journalism from around the world. Submission can be by multiple channels including social media, file sharing, transfer and upload. Short videos will be prepared to help contributors submit material safely and in a usable format. What part does the roadmap play in monitoring the delivery of the solution?

   a. The roadmap shows the order in which the submission channels will be developed and released so the videos can be produced in the right order.

   b. The roadmap shows the order in which the submission channels will be developed, but there is a separate schedule for releasing the product and the associated videos.

   c. The roadmap shows the dates by which the submission channels and videos will be developed and released so that progress can be monitored.

   d. The roadmap shows the order in which the submission channels and videos will be developed, but the dates are shown on separate project plans which are used to monitor progress.

2. Citizen journalists submit reports, photos and videos 24/7/365. The channels are therefore guaranteed to have high availability for immediate use more than 90 per cent of the time and at least two channels are always available. How is this arrangement known?

   a. Key performance indicator (KPI).

   b. Service level agreement (SLA).

   c. Non-functional requirement (NFR).

   d. Decision support system (DSS).

3. To enable citizen journalists to submit reports, four high-level deliverables have been identified: submission, validation, communication and help videos. What type of model could be used to find lower-level building blocks?

   a. Building block decomposition grid.

   b. Delivery roadmap metamodel.

   c. Work breakdown structure (WBS).

   d. Cluster analysis diagram.

4. Which **two** items are inputs to the delivery roadmap?

   i. Quality assurance and test plans from the development teams.

   ii. Gap report with changes identified and itemised.

   iii. Change management policies and procedures.

   iv. Relevant dependencies from existing project plans.

   a. i and ii only.

   b. iii and iv only.

   c. i and iii only.

   d. ii and iv only.

ANSWERS TO REVIEW QUESTIONS AND POINTERS FOR ACTIVITIES

CHAPTER 1

ANSWERS TO REVIEW QUESTIONS

1. (b)   2. (c)   3. (c)   4. (a)

POINTERS FOR ACTIVITIES

Activity 1.1 (section 1.4.3)

Improve customer service:

- **Specific:** customers receive the help they need whenever they need it 24/7.
- **Measurable:** requests are acknowledged immediately with an answer or action plan, zero complaints, customer satisfaction rating from poor (current) to very good, records of all to be kept and reported to management.
- **Actionable:** systems, processes, staff, client communication details and possibly other aspects of dealing with customers can be changed.
- **Realistic:** other comparable organisations already achieve better results in this area, so it is possible; however, the changes need to be made quickly before shareholders abandon the company.
- **Time-bound:** visible improvements must be seen within four months, targets must be achieved in seven months, and a comprehensive solution must be presented to the board within two months.

Note: This could be split into multiple objectives.

Activity 1.2 (section 1.5.2)

Stakeholder list:

- The problem area is currently centred in Outpatients so the operations director would be a suitable business sponsor. Another possibility is the finance director as the solution will save costs. However, the operations director has more to lose if things go wrong.

- The medical director is in charge of medical specialties that set up and see patients in the majority of clinics and provide results. The director of nursing manages the nurses who provide services in Outpatients and there are also nurse-led clinics. These two directors will be important senior stakeholders, as will many of their staff. Representing such large numbers will require careful selection, delegation and organisation.

- The director of strategy, finance director, chair and chief executive will be involved in governance activities.

Activity 1.3 (section 1.6.1)

Business requirements:

- Highest standards of patient confidentiality must be maintained. Patient data is only visible to authorised people and for legitimate business purposes where consent has been recorded. Note that this is likely to be a policy or principle that covers all solutions.

- Messages sent to patients must be received and understood. Note that this requirement is not explicitly stated but is an important part of ensuring that patients attend appointments.

- Messages to be available in all the languages spoken by potential patients in the Fallowdale region. Note that this is not explicitly stated but is part of the vision of the hospital to 'treat every person as an individual'.

Activity 1.4 (section 1.8.5)

Components that are likely to be part of the Fallowdale Hospital patient communication solution:

- **People:** patients, Outpatient administrative staff and staff who run clinics will all be actors in the solution.

- **Organisation:** outpatient department, medical specialties and other departments that will run clinics.

- **Process:** booking, rescheduling and cancelling appointments.

- **Information:** the content of the messages, some templates, some patient data, some scheduling data.

- **Technology:** some method of delivering messages – currently printing and posting, but could include electronic delivery in the future.

Note: The above suggestions are highly likely to be part of the solution. This does not mean there will be any changes, although it is likely. The category that is least likely to change is organisation.

CHAPTER 2

ANSWERS TO REVIEW QUESTIONS

1. (b)   2. (d)   3. (a)   4. (c)

POINTERS FOR ACTIVITIES

Activity 2.1 (section 2.1)

Fallowdale Hospital enterprise principles relevant to patient communications solution:

- **Providing services to everyone, based on clinical need:** accessibility, translation, multiple communication methods.

- **Demonstrating excellence and professionalism:** messages must be clear and related to care being offered.

- **Putting the patient at the heart of everything:** could involve increased communication with patients.

- **Working across organisational boundaries:** communication with other healthcare professionals in other hospitals and in the community.

- **Providing best value for taxpayers' money:** reducing costs and increasing the quality of communication with patients.

- **Being accountable to the public and our community:** may require publishing statistics and other information.

Activity 2.2 (section 2.4.3)

Steps in the supermarket value chain that may need to change to support the elimination of non-recyclable packaging:

- Product selection (not impacted).
- Supplier selection: additional constraint on packaging (to be recyclable and marked as such with any changes to conditions for storage and transport).
- Supplier contract: additional constraint on packaging (as above).
- Call-off of supply (not impacted).
- Supply of goods: check packaging compliance on receipt.
- Warehousing: check and apply conditions for storing packaging types.
- Packaging: if done by supermarket, constraint on materials used.
- Distribution: check and apply conditions for transporting goods with new packaging.
- Display: identification of materials used and recycling status to be prominently displayed.

Activity 2.3 (section 2.5.5)

Data entities relevant to the organic fruit farm monitoring and regulation solution could be:

- **Plant:** type, location, health.
- **Food:** type, amount, date and time given.
- **Environmental factor:** type, measurement, unit.
- **Water:** date and time given, delivery mechanism.
- **Location:** type, size, soil type, GPS coordinates.

Activity 2.4 (section 2.9.4)

Infrastructure services required by the restaurant food ordering and delivery solution:

- Communication between table and kitchen.
- Transport of food and drink to table.
- Payment mechanism.

CHAPTER 3

ANSWERS TO REVIEW QUESTIONS

1. (c)   2. (d)   3. (b)   4. (a)

POINTERS FOR ACTIVITIES

Activity 3.1 (section 3.3.1)

Suggested values for quantifying the Fallowdale Hospital patient communications problem and measuring the current and future states:

- Cost of communication: by contact (each letter), by referral (patient episode), by appointment. This should be a marginal cost rather than including overhead costs.
- Number of non-attendances: total, by clinic.
- Number of undelivered letters: this probably needs to be estimated.
- Call centre costs related to patient communication.
- Overall cost of postal communication related to patient communication.
- Number of incidents attributed to patient communication failures or inefficiencies.
- Cost of incident investigations.
- Maximum clinic capacity and actual occupancy as a percentage.
- Loss of income due to low clinic occupancy.

Activity 3.2 (section 3.4.2)

Existing architecture artefacts that are relevant to the Fallowdale Hospital patient communications solution:

- **Business:** process documentation including referral management, waiting list management, call centre management and post management.
- **Data:** data structure definition for patient contact details and appointment details, letter templates.
- **Applications:** catalogue details for applications supporting letter definition, requesting, printing and reporting. Cross-reference grids showing applications used for managing outpatients from referral to discharge.

Solution stakeholders linked to the artefacts:

- Primary care physicians (GPs) who refer patients.
- Department staff who manage the waiting lists.
- Call centre staff.
- Post room staff.
- Patients.
- Clinic staff.
- Data managers.

Activity 3.3 (section 3.6.1)

Cloud-based customer relationship management (CRM) system business case option:

Description: customer contact details to be stored and managed in a cloud-based software as a service (SaaS) application provided off the shelf by a reputable supplier but requiring some configuration to the needs of the hospital. The CRM system can produce, dispatch and record communications to and from patients and make them available to hospital staff.

Gap analysis: this is a completely new component that has already been developed; however, it does require configuration to match the hospital data. There is also work needed to provide interfaces to the existing systems where the patient data is managed. Staff in several areas of the hospital will have to alter their processes to use the CRM system. It is estimated to take between 3 to 6 months' work involving a small specialist team with ad hoc involvement by many stakeholders and some dedicated work by SMEs from the affected departments.

Cost benefit analysis: there are both capital and recurring costs involved. The capital costs are the time and resources of the programme team and dedicated time of SMEs. The recurring costs are the licence for the software and associated support by hospital and supplier staff. The benefits are the achievement of a working system in a short timescale for part of the solution. The fact that this component can be run alongside the existing solution may be seen as a benefit, although it has been noted that this may limit the opportunity for closer integration in the future.

Return on investment: if 30 per cent of patient communications can be moved to non-postal methods, the cost reduction from printing and postage would cover the cost of running the new cloud-based system. This assumes there are no price rises from the supplier or increases in the need for support. We estimate this can be achieved in the first nine months of operation. The break-even point would then come at 12 to 15 months after the start of operations. This figure of 30 per cent in nine months is reasonable based on patient survey results.

Risk assessment: the supplier has given guarantees for the continuation of service but there is always a risk of failure. This risk is increased if all communication channels are moved to the cloud and reduced if only one or a few are moved. The threat of cyberattack is roughly equal to using the usual data centre, but the risk and responsibility for control and mitigation is shared with the supplier. Security controls can be increased, for example by having dedicated equipment, but the costs increase as a result. The data is classed as personally identifiable information (PII), so the supplier must ensure this is stored and used according to data protection laws.

Activity 3.4 (section 3.9)

(a) Roadmap elements for minimum set:

- Existing standard messages based on a template (changes made separately to each of 100+ templates).
- Postal service delivery channel using printed letters (currently in use).
- Email delivery channel.

(b) Elements for next update:

- Integration with waiting list management processes (including add to list, remove, prioritise and suspend).
- Integration with referral management processes (including referral received, accept and reject).
- Instant message delivery channel.

(c) Explanation of choices:

The hospital needs the existing communication service to continue with as-is messages and postal delivery of the same. However, this does not give any benefits, so moving some patients to email will provide cost savings even if not integrated with business processes.

Larger benefits will be realised through integrating with waiting list and referral management and this can be automated for some types of communication. Adding instant messaging is similar to email and has the same dependencies; this will increase benefits and reach more recipients as well as providing a basis for responses (which will be in a future release).

Other elements will be delivered progressively and the order needs to be prioritised with stakeholders.

CHAPTER 4

ANSWERS TO REVIEW QUESTIONS

1. (b)   2. (c)   3. (d)   4. (a)

POINTERS FOR ACTIVITIES

Activity 4.1 (section 4.2.3)

The problem domain is centred on how the hospital communicates with patients. Patients are customers of the hospital and so it seems logical that gaining an insight into the interaction from the customer viewpoint must be able to illuminate the problem and help to find a solution. Note that patients are not the only customer of the hospital. Community physicians (general practitioners) and others refer patients to the hospital and are therefore also customers. Any outputs from the VoC process can be used as inputs to solution architecture and should prove very useful.

There are a large number of patients, so using a survey is one possible approach. For more detailed information it would also be possible to set up a focus group, but it would have to be representative of all patient categories and this could prove unwieldy. Apart from patients, anyone referring patients, such as general practitioners, should be included. Some non-patients are involved in the communication process, such as parents and guardians of minors, and carers for elderly and vulnerable patients.

Activity 4.2 (section 4.5.3)

Business rules can be written for the three problems identified:

- **Patients not wanting emails:** messages must only be delivered by email where the recipient has consented. If consent has not been given for any communication method, a letter is to be sent by post.

- **Invitations arriving too late:** invitations must be delivered to the patient a minimum of seven days before any appointment. Each communication method has a delivery time, for example letters take two days to arrive. If no communication method will deliver the invitation to the patient in time, an alert should be raised to the Outpatients administration manager for action. (They will try to contact the patient by phone for short-notice appointments.)

- **Allocating slots:** there is probably a business rule for this activity, and the only way to find it is by interviewing the staff who currently do the allocation. Once the interview notes have been analysed, possibly by a business analyst working in this area, then a consensus business rule can be documented and agreed with all the relevant stakeholders.

Activity 4.3 (section 4.4.6)

Accessibility is a requirement for all services provided by the hospital. For the patient communication solution, the following aspects require accessible design:

- **Messages (direct):** should be written in clear, simple language. Symbols can be used if appropriate.

- **Languages (direct):** all messages must be available in all languages spoken by people in the catchment area for the hospital. The language preferences must be captured and recorded for all recipients (patients, parents and guardians of minors, carers). A mechanism for accessing the message in other languages must be available to all recipients.

- **Support for adaptations (direct):** communication methods must be able to deliver messages in adapted format such as large print, contrast print and Braille. Preferences must be recorded for all recipients.

- **Support for adaptations (indirect):** communication methods must support adaptive technology such as screen readers and Braille readers.

Activity 4.4 (section 4.7.7)

Baseline artefacts that might help with design for capturing and managing recipient preferences are:

Previous business case(s): if a similar solution has been considered before, then looking at the business case would be useful. It seems unlikely in this case.

Baseline solution architecture artefacts: again, this seems unlikely. The current system of sending letters is very old.

Enterprise data models: these are likely to be helpful – for example, models of data that is currently stored about patients. This could help with deciding the best place in the data model to add details of preferences. How parents, guardians and carers are linked to patients is also relevant. A data model for the message content would also be useful, since this will have to be modified in line with stated preferences.

Applications catalogue: this may show how patient and message data is currently being managed.

Business model: the business model does not currently include any activities that use personalised communication preferences, but it is possible that the patient communications solution changes the business model, in which case it would need to be updated.

Business service catalogue: this is worth looking at, to see if any other business services deal with other languages and adaptations.

Infrastructure models: solution architecture designs are at the logical level, so these are unlikely to be useful.

CHAPTER 5

ANSWERS TO REVIEW QUESTIONS

1. (a)   2. (c)   3. (d)   4. (b)

POINTERS FOR ACTIVITIES

Activity 5.1 (section 5.2.1)

Analysis of components and interfaces for business case option 2:

- **Clinic administrator (person):** modified role, increased responsibilities affecting capacity, training required, new interface (2) with CRM system.
- **Clinic management (system):** new interface (1) providing clinic booking information to CRM, removed interface U to letter printer and folder.
- **CRM system (system):** new component with interface (3) to letter printer and folder (with same message format as previously sent from clinic management), interface (4) to text system and interface (5) to email system.
- **Letter printer and folder (system):** reduced capacity in future, print instructions to be received from CRM system in the same format as previously from clinic management.
- **Postal service (service):** unchanged with capacity reduced in future.
- **Patient (person):** possible new communication methods (email and text); these changes need to be notified to patients and consent obtained to use them.
- **Text system (system):** new component with interface (6) with patient.
- **Email system (system):** new component with interface (7) with patient.

Additional details include any costs that are known about, an estimate of the effort and non-financial resources required, and an estimated time to complete the change.

Activity 5.2 (section 5.3.2)

Of the four existing components identified:

Each of these is only relevant to gap analysis if the target AD contains a component that requires the functionality provided.

For business case option selection, for example, all of the proposed options have an email component and two of the three have a text component. None has a social media component, so this does not seem relevant. The models in the business case are too high-level for any meaningful discussion about involving the PR unit in composing patient messages.

For the main gap analysis, which is performed at the end of the validation phase, the existing email component will be relevant, but more information is required to see if it can be part of the solution. For example, can all the required functionalities be provided and can it scale up to the required level?

The text system could be relevant, but only if the final solution design has a text component. If so, similar questions about functionality and scalability would need to be asked. Even if not directly relevant to the gap analysis, information could be included as an appendix to the gap report as it is a likely candidate for a future target architecture.

The existing social media presence is interesting but not directly relevant. This may also be mentioned in an appendix.

The PR business unit component is also of interest as it could become a component in the solution. During the logical design phase, the architecture team will have to decide how new messages will be created and existing ones maintained. Although the initial solution is likely to be limited to outpatient appointment invitations, even these contain hospital messages and informative enclosures. It is highly likely that a department such as the PR unit has some input and possibly responsibility for the composition of these messages.

Activity 5.3 (section 5.4.5)

Figure 5.12 shows the gap analysis result using the graph method.

Figure 5.12 Gap analysis graph

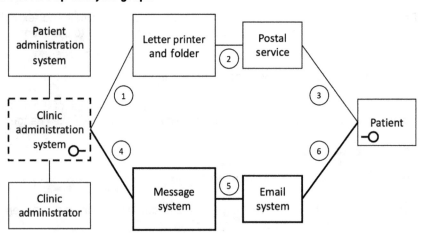

CHAPTER 6

ANSWERS TO REVIEW QUESTIONS

1. (d)   2. (c)   3. (a)   4. (b)

POINTERS FOR ACTIVITIES

Activity 6.1 (section 6.3.1)

Components that are relevant to the cost of the solution are:

- **Clinic management (person):** possible change in capacity, fixed cost of staff.
- **Letter printer and folder (system):** likely reduction in capacity, fixed cost of lease purchase, marginal cost of consumables.
- **Post room staff (organisation unit):** likely reduction in capacity, fixed cost of staff.
- **Email system (system):** likely increase in capacity, fixed cost of system management.
- **Email system, message content and mail merge components (system):** initial cost of data transfer, ongoing fixed cost of updating message content.

Note that most of the costs associated with sending letters are fixed, with only consumables being variable. This means that keeping the existing systems and staff in place will not reduce costs and there are some additional costs with other means of communication such as email.

Suitable visualisations could include a table of data and a bar chart that would allow the FD to see the details of the costs, before and after the new solution is deployed.

Activity 6.2 (section 6.5.1)

Apart from the print room, other business components in scope for the Fallowdale Hospital patient communications solution include:

- **Outpatient administration (organisation):** outpatients manager.
- **Medical departments that run clinics (organisation):** a representative of one or more departments.
- **Postal service (organisation, external):** service manager from the postal service.
- **Public relations (organisation):** PR manager.
- **Book appointment (process):** outpatients manager.
- **Other outpatient processes, such as cancel and reschedule (process):** outpatients manager.
- **Clinic management (job role):** outpatients manager.

Activity 6.3 (section 6.7)

Possible stakeholders for the solution design authority include:

- Outpatients manager.
- Print room manager.
- Representative of a medical department.
- Member of the solution architecture team to organise, present designs and communicate decisions.

Other stakeholders can be asked for expert input when necessary. Decisions can be escalated to the operations director (business sponsor) when necessary.

Activity 6.4 (section 6.7)

Stakeholders to be invited by the solution design authority to provide expertise for the first meeting include:

- **Changes to the letter printer and folder:** the supplier can advise on changes to the contract.
- **Safeguarding patient information:** a security architect or security specialist.
- **Obtaining patient consent:** PR manager, a specialist in customer experience management, or someone involved in an existing solution that requires patient consent to be recorded.
- **Controlling the content of messages:** business analyst or other role with good understanding of processes involving communication management.

CHAPTER 7

ANSWERS TO REVIEW QUESTIONS

1. (b)   2. (a)   3. (c)   4. (d)

POINTERS FOR ACTIVITIES

Activity 7.1 (section 7.3)

Possible baseline opportunities include:

- **Automation:** use the RPA scheduling tool to allocate clinic slots and possibly initiate communication. This would expand the scope of the solution and the risks would have to be assessed, but this could reduce the ongoing costs of the solution and improve performance and reliability.

- **Communication:** the use of the chatbot is an attractive idea but may require an expansion of the scope of the solution; currently the only interaction with the patient is to get and maintain their communication preferences. This would be a good component for dealing with confirming, cancelling or rescheduling appointments, especially out of hours.

- **Cloud:** making new components cloud-native seems to be a priority, development aspects are likely to be more relevant to the completion phase of the solution architecture life cycle; the infrastructure architecture function should be aware of any overlap between the solution's infrastructure service requirements and plans to move services to the cloud. It would be worth considering new components such as sending emails or logging patient communication to see whether cost savings and other efficiencies can be achieved by moving existing services to the cloud.

- **Data:** analytics are likely to be useful in spotting patterns of patient behaviour such as the response to different communication media. This could lead to service improvements and should be considered either now or for the future when more data is available. Flexible storage alternatives to the highly structured relational database should be considered, such as document or tabular databases that may be better suited to storing patient communication preferences that have a potentially unlimited number of options, with a variety of parameters.

Activity 7.2 (section 7.4)

Infrastructure services to support remote appointments could include:

- **Client access:** website, virtual client for patients to connect to a remote appointment.
- **Data storage:** NoSQL to record appointments for safeguarding or auditing, storage of preferences for appointments.
- **Networking:** capacity may need to be improved for audio and video connections.
- **Communication:** video and voice facilities.
- **Security:** encryption, identification of patients.

Activity 7.3 (section 7.4)

SBBs may be classified and associated with infrastructure service requirements as follows:

- **Communications management system (technology):** flexible data storage, message sending and receiving, logging.
- **Email generator (technology):** email communications, data validation, logging.
- **Letter printer (technology):** printing, folding and enveloping, data validation, logging.
- **Manage waiting lists (process):** automation (perhaps by robotic process automation), data validation, message sending (for escalation), client access devices.

- **Medical or diagnostic unit (organisational unit):** client access devices, data entry, management information (MI) and possibly data analytics to forecast demand.

- **Message contents (information):** data storage, data validation, integration to amalgamate message contents from multiple sources.

- **Patient communication preferences (information):** flexible data storage, data validation (for currency).

- **Patient record (information):** structured data storage, data validation, data verification, encryption and other security controls appropriate in line with enterprise directives for highly confidential information.

- **Post room (organisational unit):** message receipt to alert for various events to do with the letter printer, client access devices to receive alerts.

- **Reschedule appointment (process):** identity management, data validation, client access devices, message sending to medical or diagnostic departments.

Note that all of the above SBBs have one or more SBBIs that will require networking services.

Activity 7.4 (section 7.6)

Options for strengthening the design for data integrity and currency include:

- Locating the communications management system (CMS) with the data in the data centre: this would lead to an increase in traffic across the link.

- Copying patient address and preference data for each clinic booking so that only one data set is required: this would lead to data duplication and risk a loss of currency and integrity.

- Add a second data connection between the hospital and the data centre: this would be an expensive option.

- Migrating the data stores and the CMS to the cloud: this would raise security concerns about patient data.

- Migrating the CMS to the cloud and connecting it to the data centre via a private network: slightly better than the previous option, but still has an external link to the data centre that is probably not allowed by the hospital's security directives.

- Pre-processing the data in the data centre to prepare a single data set containing all the relevant data: this would make better use of the data connection and reduce the risk of failure, but does require an extra processing service at the data centre; however, it is the best option.

Activity 7.5 (section 7.7)

Recommendations for allowing the patient to update their communication preferences securely include:

- Securing the website using encryption so any credentials cannot be intercepted.

- Using identity and access management (IAM) to identify the patient in a secure fashion: an IAM component is available in the infrastructure technology catalogue.

- Defining a process for patient registration and access that meets the requirements of ISO 27001.

- Logging all successful and unsuccessful attempts to log in and update data and provide escalation alerts for all suspicious activity.

- Ensuring only the minimum information necessary is visible on the website.

CHAPTER 8

ANSWERS TO REVIEW QUESTIONS

1. (c)   2. (b)   3. (c)   4. (d)

POINTERS FOR ACTIVITIES

Activity 8.1 (section 8.1.4)

Possible components and supporting activities:

- Preference data storage.

- Collect preference and consent data from patients.

- Collect email and text contact details from patients.

- Set preference to letter by default until preferences are received.

- Preference management system for patients/staff to update preferences, addresses and renewing consent.

Order of delivery:

- **1:** Storage required before anything can be set.

- **4:** To enable letters to continue.

- **3:** Has a dependency on the contact details (this is under 'Address' on the WBS).

- **2:** Could be combined with 3 to minimise contact with patients.

- **5:** Required for ongoing maintenance of data, initial setup probably more manual.

Activity 8.2 (section 8.2.4)

Early delivery of logging system:

- Could mean early delivery of legal auditing and MI.

- Options are to update roadmap or leave delivery schedule unchanged.

- Roadmap changes need to be escalated to business owners and senior managers.
- If a decision to change the roadmap is made and approved by the design authority, the business sponsor will need to authorise P3M to make the necessary changes to all affected programme and project plans.

Late delivery of the MCMS:

- Without the MCMS, message content cannot be kept up to date and this would mean the solution fails to deliver some key business requirements.
- The decision is whether to use the work-around or delay other deliverables.
- Using the work-around would not require changing the roadmap.
- Delaying other deliverables would require changing the roadmap.
- The use of back-office IT staff would increase the cost of the solution.
- Roadmap changes need to be escalated to business owners and senior managers.
- Decisions that would increase costs need to be escalated to the business sponsor.
- Whichever decision is made by the design authority, the business sponsor will need to authorise P3M to make the necessary changes to all affected programme and project plans.

Activity 8.3 (section 8.4)

This table shows suggested service levels for each capability in the roadmap:

Capability	Service level	Tolerances and escalation	Measure and frequency
Ad hoc email	Availability 7am to 7pm	> 5 minutes escalate to service management and alert to management	Server heartbeat, 1 minute
	Email address verified 100%	No response 5 days after consent expiry escalate to service management, add to weekly MI report	Currency of record, 1 day
Legal auditing	Response to request for information < 4 hours including verification	Escalate to management, alert to requester and service management, national regulator has 7-day statutory limit so > 24 hours escalate to board	Request log, 1 hour
MI	On schedule, usually daily, weekly or monthly	Delay > 1 hour escalated to service management and alerted to recipient	Schedule vs log, 1 hour
	New requests < 24 hours	> 1 hour escalate to service management, alert to recipient	Request log, 1 hour

Capability	Service level	Tolerances and escalation	Measure and frequency
Patient responses	Acknowledge < 1 hour	> 30 minutes escalate to service management	Response time, 1 hour
	Resolve < 24 hours	Delay > 1 hour escalated to management	Response time, 1 hour
Structured messages	Compliance with data model for integrity 100%	Escalate to data management and alert management	Validation against data model, 1 day
	Availability 24x7	> 5 minutes escalate to service management, alert management	Server heartbeat, 1 minute
	Response time < 1 minute	> 5 minutes, escalate to service management, alert management	Response log, 1 minute
Auto invitations	Dispatch < 30 minutes from trigger request	Dispatch > 1 hour, escalate to service manager and alert management	Message log, 15 minutes
	Zero rejected messages	Escalate to management, notify data management	Message log, 15 minutes
Data analytics	This is a complex and varied service, a bit like software development. Established analytics have similar service levels to MI. New projects are usually based on an estimate.		

Activity 8.4 (section 8.5)

Modified components:

- Patient preference data to include phone and video.
- Patient contact details to include phone and video.
- Patient consent to include recording phone and video.
- Consent and preference.
- Clinic administration system updates for phone and video appointments including slot types and attendance recording.
- Message contents new instructions for phone and video appointments, including letter templates in short term.
- Network capacity increase.

Additional components:

- Video communication system.
- Phone communication system.

- Equipment in clinic.
- Technical support for clinic.
- Technical support for patient.
- Storage and retrieval of recordings.
- Patient equipment such as phone, computer, video camera.

Impact on service levels:

- Possible increase in overall number of appointments and therefore invitations and responses.

Additional impacts:

- Increased attendance.
- Increased acceptance of first offered appointment.
- Increased flexibility for offering appointments, e.g. a couple of slots between other jobs.
- Increased pressure on community (GP) diagnostics, e.g. phlebotomy, to support remote appointments.

REFERENCES

Alexander, C., Ishikawa, S., Silverstein, M., Jacobson, M., Fiksdahl-King, I. and Angel, S. (1977) *A Pattern Language: Towns, Buildings, Construction*. New York: Oxford University Press.

BCS (2019) Foundation Certificate in Business Change Syllabus. Available from: www2.bcs.org/certifications/ba/business-change-foundation-certificate.

Business Architecture Guild (2021) Information about the Business Architecture Guild. Available from: businessarchitectureguild.org.

Cadle, J. and Paul, D. (eds) (2021) *Business Analysis Techniques: 123 Essential Tools for Success*, 3rd edition. Swindon: BCS.

Cambridge Dictionary (2021) 'Strategy'. Cambridge: Cambridge University Press.

Data Protection Act (2018) *Data Protection Act 2018*, Ch. 12. London: United Kingdom Government.

DoDAF 2.02 (2010) *Department of Defense Architecture Framework Version 2.02*. USA: U.S. Department of Defense.

Doran, G. T. (1981) There's a S.M.A.R.T. way to write management's goals and objectives. *Management Review*, 70 (11). 35–36.

DSDM Agile Business Consortium (2014). Agile Project Framework Handbook. Available from: agilebusiness.org.

Formal/2015-05-19 (2015) *Business Motivation Model Version 1.3*. Milford, MA: Object Management Group.

Formal/2017-12-05 (2017). *OMG® Unified Modeling Language® Version 2.5.1*. Milford, MA: Object Management Group.

Gamma, E., Helm, R., Johnson, R. and Vlissides, J. M. (1994) *Design Patterns: Elements of Reusable Object-Oriented Software*. Reading, MA: Addison-Wesley.

Griffin, A. and Hauser, J. R. (1993) The voice of the customer. *Marketing Science*, 12 (1), 1–27.

Harry, M. J. (1988) *The Nature of Six Sigma Quality*. Rolling Meadows, IL: Motorola University Press.

ISO/IEC/IEEE 42010:2011 (2011) Systems and Software Engineering – Architecture Description. Switzerland: International Organization for Standardization.

ISO/IEC 27001:2013 (2013) Information Technology – Security Techniques – Information Security Management Systems – Requirements. Switzerland: International Organization for Standardization.

ISO 27002:2013 (2013) Information Technology – Security Techniques – Code of Practice for Information Security Controls. Switzerland: International Organization for Standardization.

ISO/IEC 2382:2015 (2015) ISO IT Vocabulary. Switzerland: International Organization for Standardization.

ISO 9241-11:2018 (2018) Ergonomics of Human–System Interaction. Part 11: Usability: Definitions and Concepts. Switzerland: International Organization for Standardization.

ISO/IEC 27000:2018 (2018) Information Technology – Security Techniques – Information Security Management Systems – Overview and Vocabulary. Switzerland: International Organization for Standardization.

ISO 50001:2018 (2018) Energy Management Systems – Requirements with Guidance for Use. Switzerland: International Organization for Standardization.

ISO 22301:2019 (2019) Security and Resilience – Business Continuity Management Systems – Requirements. Switzerland: International Organization for Standardization.

Jacka, J. M. and Keller, P. J. (2009) *Business Process Mapping: Improving Customer Satisfaction.* Hoboken, NJ: Wiley.

Kotonya, G. and Somerville, I. (1998) *Requirements Engineering: Processes and Techniques.* London: Wiley.

MITRE Corporation (2021) Common Weaknesses Enumeration (CWE). Available from: cwe.mitre.org.

Murray-Webster, R. and Dalcher, D. (eds) (2019) *APM Body of Knowledge.* 7th edition. Princes Risborough: Association for Project Management.

Nielsen, J. (1993) *Usability Engineering.* San Francisco, CA: Morgan Kaufmann.

Ohno, T. (1988) *Toyota Production System: Beyond Large-scale Production.* New York: Productivity Press.

Osterwalder, A. and Pigneur, Y. (2010) *Business Model Generation: A Handbook for Visionaries, Game Changers, and Challengers.* Hoboken, NJ: Wiley.

OWASP (2021) OWASP Top Ten. The Open Web Application Security Project.® Available from: owasp.org/www-project-top-ten.

Oxford Dictionary (2021). 'Stakeholder'. Oxford: Oxford University Press.

Paul, D. and Cadle, J. (eds) (2020) *Business Analysis*, 4th edition. Swindon: BCS.

Porter, M. E. (1979) How competitive forces shape strategy. *Harvard Business Review*, March 1979.

Porter, M. E. (1980) *Competitive Strategy.* New York: Free Press.

Porter, M. E. (1985) *Competitive Advantage: Creating and Sustaining Superior Performance.* New York: Simon and Schuster.

Regulation (EU) 2016/679 (2016) *General Data Protection Regulations.* Brussels: European Union.

Schwaber, K. and Sutherland, J. (2020) The Scrum Guide. Available from: scrumguides.org.

Sondhi, R. (1999) *Total Strategy*. Lancashire: Airworthy Publications.

Sutcliffe, A. (1996) A conceptual framework for requirements engineering. *Requirements Engineering*, 1. 170–189.

Tannady, H., Andry, J. F., Gunawan, F. E. and Mayseleste, J. (2020) Enterprise architecture artifacts enablers for IT strategy and business alignment in forwarding services. *International Journal of Advanced Trends in Computer Science and Engineering*, 9 (2), 1465–1472.

TOGAF 9.2 (2018) *The TOGAF® Standard, Version 9.2*. San Francisco, CA: The Open Group.

World Wildlife Fund (2021) Mission Statement. Available from: wwf.org.uk/who-we-are.

Zachman, J. A. (2008) The Concise Definition of the Zachman Framework. Available from: zachman.com/about-the-zachman-framework.

FURTHER READING

Project management of solutions

Blokdyk, G. (2021) *Solution Architecture: A Complete Guide*. Brisbane: Emereo Publishing.

McSweeney, A. (2019) *Introduction to Solution Architecture*. Independently published.

Solution architecture specification

Gu, S. (2021) *Agile Enterprise Solution Architecture: An IT Service-Based Modeling Approach*. Colorado Springs, CO: Vernal Press.

Solution architecture with specific technologies

Ramirez, G. (2018) *AWS Certified Solutions Architect – Associate Guide: The Ultimate Exam Guide to AWS Solutions*. Birmingham: Packt Publishing.

Shrivastava, S. and Srivastav, N. (2020) *Solution Architect's Handbook*. Birmingham: Packt Publishing.

Enterprise architecture

Ross, J. W., Weill, P. and Robertson, D. (2006) *Enterprise Architecture as Strategy: Creating a Foundation for Business Execution*. Cambridge, MA: Harvard Business Review Press.

Web resources

Solution architecture skills specification. Available from: sfia-online.org/en/sfia-7/skills/solution-architecture.

Solution architect: Processes, role description, responsibilities, and certifications. Available from: altexsoft.com/blog/engineering/solution-architect-role/.

Comparison between the roles of enterprise, solution and technical architects. Available from: leanix.net/en/enterprise-architect-vs-solution-architect-vs-technical-architect-whats-the-difference.

INDEX

Figures and tables are given in italics.

www.ingramcontent.com/pod-product-compliance
Lightning Source LLC
Chambersburg PA
CBHW060531060326
40690CB00017B/3449